Political Economy of Steel Development in Nigeria: Lessons from South Korea

Daniel Omoweh

Africa World Press, Inc.

P.O. Box 1892
Trenton, NJ 08607

P.O. Box 48
Asmara, ERITREA

Africa World Press, Inc.

P.O. Box 1892
Trenton, NJ 08607

P.O. Box 48
Asmara, ERITREA

Copyright © 2005 Daniel Omoweh
First printing 2005

All rights reserved. No part of this publication may be reproduced, stored in a retrieval system or transmitted in any form or by any means electronic, mechanical, photocopying, recording or otherwise without the prior written permission of the publisher.

Book Design: Jonathan Gullery
Cover Design: Ashraful Haque

Library of Congress Cataloging-in-Publication Data

Omoweh, Daniel A.
  Politcal economy of steel development in Nigeria: lessons from South Korea / by Daniel A. Omoweh.
    p. cm.
  Includes bibliographical references and index.
  ISBN 0-86543-914-1 (hardcover) -- ISBN 0-86543-915-X (pbk.)
  1. Steel industry and trade--Nigeria. 2. Steel industry and trade--Korea (South). I. Title.

HD9527.N52 O46 2001
338.4'7669142'09669--dc22

2001005351

Table of Contents

Preface and Acknowledgements 5

Chapter One: Introduction to the Study 9

Chapter Two: Contending Theoretical Issues 17

Chapter Three: The Politics of Steel Development
                   and Industrialisation 53

Chapter Four: Structure of the Steel Industry 91

Chapter Five: Performance of the Steel Industry .. 125

Chapter Six: Problems and Prospects 159

Chapter Seven: Conclusion:
                   Policy Recommendations 195

Bibliography 217

Statistical Appendix: Tables and Figure 226

Index   261

Preface and Acknowledgement

This study of Nigeria's steel sub-sector role in its quest for industrialisation was undertaken as part of my research project at the Nigerian Institute of International Affairs, NIIA, Lagos. Its primary object is to review Nigeria's experience with industrialisation and steel development in comparison with South Korea's since the end of the World War 11 in 1945, with the hope of identifying the problems and prospects of the steel sector of both countries. Within this context, the study described extensively, the origin, nature, growth and problems of the steel industry with emphasis on the intermediate and capital goods-producing companies in Nigeria and Korea. Taking into account Korea's relatively more impressive growth in its steel sector notwithstanding the country's poverty in steel raw materials, this study has, within this framework too, critically examined the specificities of the industrial and steel policies of Korea, taking cognizance of their strengths and weaknesses, and the policy implications that all this portends for Nigeria.

This study is structured into seven chapters. Chapter One introduces the work, and defines its scope and methodology. Chapter Two sets the theoretical context of the study as well as providing a rigorous basis for comparing the experiences of Nigeria and South Korea with industrialisation. Since the state is still at the heart of the steel industry in both Nigeria and South Korea, Chapter Three takes a comparative and critical look at the nature of the state and its mode of extracting surplus from the economy and how all this has come to shape its role in developing the steel sector in particular, and industrialisation, generally. Chapter Four critically examines the structure of the steel industries of both countries. Chapter Five evaluates the performance of the steel companies in the contexts of the factors militating against their operations, and their contribution to the production of intermediate and capital goods, and technological development. Chapter Six analyses the nature of the problems facing the iron and steel companies of both Nigeria and South Korea and the prospects of their survival. Chapter Seven concludes the study with policy prescriptions.

I would like to first thank God for His infinite mercy on me, particularly for sustaining me in the course of this project. I owe a huge debt of gratitude to Josephine, my wife, and Mudia, my daughter, who all endured my absence from home when this study was undertaken. I would also like to thank George Obiozor and the rest of my colleagues in the Research Department for their moral support. In particular, I am eternally grateful to R.A.Akindele, who readily rendered me valuable advice on academic research; and to A.O.Olukoshi, who encouraged me to research on Nigeria's industrialisation. The staff of the Institute's library also helped me a great deal with the literature in the field of study. For all this, I remain grateful. I would like to thank M.O.Osaweme for his moral support. My special thank goes to Julius Ihonvbere of the Ford Foundation, New York, for his constant scholarly advice since 1984 when I first met him at the University of Ife.

The Korean component of this study was made possible by the Korean Foundation under whose fellowship I undertook a six-month research trip to South Korea. I would like to specially thank the Korean government and the Foundation for sponsoring my research in Korea. In particular, I express my gratitude to the President and the entire staff of the Foundation for their hospitality during my stay in Korea. Carrying out this study in Korea would have been difficult as my previous letters to three institutions in Korea to which I wanted to affiliate with, were all turned down basically for reasons of lack of office space and accommodation. But within one week of his receipt of my letter of affiliation to Donggkuk University, Sing Young LEE, a Professor of International Trade in the Department of International Trade of the same university, secured a position for me at its [Donggkuk University] Institute of Management and International Trade, Seoul. His hospitality to me was unequalled. For all this, I am very grateful. I would also like to express my gratitude to the President and entire staff of Donggkuk University for their cooperation during my stay in Korea.

I just hope that, at the end of the day, this study would have met the objectives it has set out to accomplish. Whatever shortcomings it contains are entirely mine.

Daniel A. Omoweh
Nigerian Institute of International Affairs
Lagos, Nigeria, May 2004.

Chapter One

Introduction to the Study

1.1 INTRODUCTION: AN OVERVIEW OF STEEL DEVELOPMENT AND INDUSTRIALIZATION IN NIGERIA AND SOUTH KOREA

Are there lessons Nigeria can learn from South Korea's experience with steel development and industrialization? This is one fundamental question that still agitates the minds of scholars, policymakers and all those who are concerned with overturning Nigeria's depeening industrial underdevelopment. The need for this comparison stems largely from the fact that, whereas Nigeria is so endowed with iron ore, coal, alloying minerals and natural gas, yet , it is unable to get its steel industry really underway, South Korea's poverty in iron ore, coal and other steel raw materials did not hinder it from developing a relatively more successful steel sector which helped launch the country's industrial age. This feat by the Korean government particularly at a time when the majority of the industrialized countries of the North are opposed to the developing economies' quest for industrialization, has endeared its policies and strategies for industrialization to some policy makers, scholars and government officials in Nigeria and other African countries. Even the international development agencies and multilateral financial institutions like the World Bank and the International Monetary Fund with policies that placed obstacles on South Korea's path to industrialization and steel development in the past, are today, hailing its efforts.

Evaluating the experiences of both Nigeria and South Korea with industrialization and steel development in particular, however, is ridden with difficulties which normally inhere in making comparisons across different regions, over different time frames, and different historical experiences and internal conditions. These difficulties notwithstanding, this study still attempts to document, describe and critically analyse in a holistic and dialectical manner, the experiences of both Nigeria and South Korea with iron and steel development, taking cognizance of the role the governments had played and still plays, in the formulation and implementation of industrial policies including steel policies, and the activities of both local and foreign steel capitals in the steel sectors of the two countries. As shall soon be made clear in the course of the discussion, the path which the governments of Nigeria and South Korea took to steel development was to a large extent, both the outcome of their historical experiences and the conception of industrialization by those in charge of the state. It is, indeed, these issues that have been at the heart of the critical emergent comparative theoretical debate on industrialization in Africa and the Asian NICs of which South Korea is one.

Since the turn of the 1960s, there has been a raging debate as to whether Nigeria, among other African countries, actually needed to establish its own iron and steel industries, or to continue to depend on Europe for its steel needs. As a newly politically independent country in 1960 with agriculture accounting for the bulk of its economic activities, the contention of some scholars then, was that, industrialization as a project, was not really a priority for those entrusted with the management of the Nigerian state. Perhaps, this explained why the development of iron and steel was not given any serious thought by the state; hence, its preference to rely on import for the country's steel needs. Little wonder why the large deposits of steel raw materials notably iron ore, cokable coal, limestone and ferro-alloys like manganese, colbalt and chromium which abound in the country, were hardly mined. And for the few minerals that were developed by the foreign companies such as coal, it was exported ostensibly to cater for the needs of the metropolitan factories. It was all the moreso since the Nigerian government, like the governments of other African countries, considered mining a risky venture and therefore, was satisfied with the rents which came mostly in the form of royalties and taxes paid by the operating foreign mining companies. Even when the state established its own mining companies like the Nigerian Mining Corporation, Jos; and the Nigerian Coal Corporation, Enugu, to exploit solid minerals, its ultimate aim was also to export since there were hardly local companies that consumed them.

By the early 1970s, foreign capital had, in its attempt to have a firmer grip of the post-colonial Nigerian economy, fully shifted focus from purely exchange to the local reproduction of hitherto imported light and intermediate goods. One major issue that arose from import reproduction was that technology still remained foreign. In essence, for the import substitution of the intermediate and capital goods to really take place in Nigeria, the technological capacity and the inputs to produce most of the machines and their parts, would have to be developed locally. As a result, the thrust of the debate on Nigeria's industrialisation in the rest of the 1970s, centered around the urgent call for Nigeria to establish its own iron and steel companies if it, indeed, wanted to industrialize. This view gained much ground among policy makers, scholars, and industrialists in the country throughout the 1980s.[1]

At the moment, the debate is no longer concerned with the need for an iron and steel industry in Nigeria as the state has already established integrated steel companies and rolling mills in the country. Rather, the question is: Why have the steel companies failed to reverse the country's industrial underdevelopment since they came on stream? At issue, is the concern for why all the public iron and steel companies which have been set up are, at their various stages of development, faced with severe problems such as acute financial crisis, cannibalization of most machines to keep few of the production lines of the steel companies functioning, prolonged shut-down period, while their counterparts in the Asian NICs which were built about the same period, have functioned relatively well and even better as well as helping to launch the industrial and technological age of these countries. South Korea, for instance, which has none of the basic steel raw materials like iron ore, coking coal and other ferro-alloy minerals; yet, it has developed a strong and flourishing steel sector compared with Nigeria. Also, there is the contention that, though British colonialism had laid the basis of Nigeria's industrial backwardness of today, Korea's relatively well developed steel industry and an overall impressive industrial growth inspite of having suffered Japan's colonial rule, is all indicative of the fact that a lot is wrong with the Nigerian state and its approach to industrialization. The state is indicted for its inability to deliver, particularly its inappropriate micro and macro-economic policies and a hostile political environment all of which have suffocated attempts at steel development and industrialization. Since the Nigerian government embraced the World Bank/International Monetary Fund-led adjustment package in 1986, and even after its official termination in 1994 since it compounded the economic problems it was planned to redress, the argument still rages on as to whether the state has really any business with iron and steel development.

Two contrasting views on these issues have emerged from the discourse so far. First, is the contention by a group of scholars, policy makers and some fraction of local capital that, since steel development requires huge capital outlay and serves as a bedrock for industrialization, the state needs to be at the centre of such a strategic sector at the early stages of its development for the purpose of protecting the country's national interest in industrial development. The relative success of Britain, France, and Germany among other European countries with steel development, and recently that of South Korea, were all cited to support such position.

Opposed to this view, are the advocates of the market approach, whose basic contention is that, since steel development is both capital and technologically intensive, it needs an efficient management system, which private capital, perhaps, has the capacity to handle. In other words, if the state has to play any role at all in the steel sector, it should be a regulatory one. Notable advocates of the market approach, are the World Bank and the International Monetary Fund. As it shall soon be made clear in the course of the discussion, the position of both the Bank and the Fund on steel development is not really surprising as they have, at different times in the past, frustrated steel projects in Nigeria and South Korea. It is not really because the state is incapable, but rather, their real intention was to protect the local steel market for the metropolitan steel companies.[2]

1.2 MOTIVES AND OBJECTIVE OF THE STUDY

This study of Nigeria's experiences with steel development and industrialization is primarily motivated by a number of factors. First, contrary to the expected role that the steel sector ought to play as Nigeria's basis of industrialization, it is still largely an enclave unto itself and increasingly becoming incapable of overturning the country's industrial underdevelopment long after attaining political independence in 1960. Worst still, the local steel companies are unable to meet even a quarter of the country's per capita steel consumption estimated at 30kg, which is far below the world approved minimum of 100kg. Second, Nigeria's industrial backwardness is deepening, with little or no hope of reversing it in sight. The country's capital goods-producing companies are still 97 percent dependent on imports for their basic production inputs. The general structure of manufacturing is, on average, an assembly of imported completely knocked down parts, CKD, which, in turn, has stifled preliminary efforts that were made by some cottage factories to indigenize technology. For, the success of the efforts of the cottage factories at replicating and accummulating technology would have helped considerably, in duplicating some of the major parts of machineries hitherto imported as well as embarking on the manufacturing of new ones. Third and final, drawing on the relatively impressive growth of South Korea's iron and steel sector, particularly its role in the country's industrialization, the study is compelled by the need to deeply understand the micro and macro-economic policy mixes and the political environment that helped shape the path which the Korean state took to industrialization. All this is in an attempt to identify possible policy errors and other factors which have militated against the development of the steel industry in Nigeria, and how it impeded its industrialisation.

As its major objective therefore, this study critically analyses the nature and process of the development of Nigeria's steel sector and industrialization, pointing out its problems and charting a new course for its survival. Within this broad objective, it examines in a holistic and dialectical manner, the origin, nature and scope of the political, economic and social forces and how their interplay have shaped iron and steel development and industrialization in Nigeria. In doing this, it also takes critical note of the structure of the steel companies, their product profile, linkages with other sectors of the Nigerian economy and the world steel industry at large. It compares and contrasts all this with South Korea's experience with steel development and industrialization, noting its industrial and steel policies, their strengths and weakness, and what it all portends for policy recommendations for Nigeria..

1.3 Context of this Study

The context of this study on Nigeria's quest for steel and industrial development in comparison with South Korea's, is in two but inter-related folds. First, it is part of my larger research agenda of comparing industrialisation in Africa and Asia. Second, as part of my research project on Nigeria's capital goods industry at the NIIA. Their inter-relatedness lies in the common objective of how this project could help deepen an understanding of the basic issues that underpin critical aspects of the international economic relations between Nigeria and South Korea in particular, and Afro-Asian studies at large. Apart from Nigeria's large population of over 100million and of course, its huge market potentials, the country is endowed with virtually all the steel raw materials in addition to petroleum most of which South Korea lacks. Furthermore, Nigeria's adoption of the path which the majority of Western European countries took to steel development has turned out a monumental failure as its steel industry is still far from really getting underway. Which means that, if South Korea lacking the basic steel raw materials has relatively done well in steel development in the past twenty years, then, its experience might contain some fundamental policy lessons that could be of interest to policy makers, government officials and local private capital in Nigeria, and other African countries. As it shall soon be demonstrated in the course of the discussion, this is by no means an assumption that, the approach that the Korean state took to the development of its iron and steel sector should be a model for Nigeria since its [Korea] experience is, from all indications, far from being one. Rather, this study's major concern is to critically examine the nature and role of the Korean state in the country's quest for industrialisation, particularly its industrial policy and the institutions established to monitor steel development; and the policy implications of all of this for Nigeria.

1.4 Notes on Methodological Issues

This study of Nigeria's experience with steel development and industrialization laid great emphasis on the generation of independent empirical data to demonstrate the specificities of the country's nature and process of industrial backwardness.

Primary data were obtained basically from interviews, oral history, newspapers, and plant tours during the various field trips which the author undertook in both Nigeria and South Korea. Empirical data were then supplemented with materials obtained from secondary sources most of which were books, journals, dissertation abstracts and relevant statistical bulletins in the two countries of study. In the presentation and assessment of data in this study, a combination of both qualitative and quantitative methods of analyses were employed.

In addition to interviews conducted with relevant public and private offficials and extensive desk research works carried out in both Nigeria and South Korea, this study also drew significantly on materials from other sources, notably government and private institutions in both countries. In Nigeria, these included the Federal Ministries of Power and Steel, Solid Minerals, Petroleum Resources, Industry, Trade and Commerce, the Central Bank of Nigeria, Nigerian Industrial Development Bank, Nigerian Bank for Commerce and Industry, the Nigerian National Petroleum Corporation, the Nigerian Mining Corporation, Manufacturers' Association of Nigeria, Federal Office of Statistcs, Federal Institute of Industrial Research, African Iron and Steel Association Secretariat, the National Association of Chambers of Commerce, Industry, Mines and Agriculture. In South Korea, data were obtained from both public and private establishments notable among which were: the Ministries of Trade and Economics, Korea Development Bank, Pohang Steel Company, POSCO Research Institute, the Libraries of the Universities of Dongguk, Korea and Seoul National. The Korea Development Institute and the Korea Steel Association were also consulted. Data were also extracted during the author's tour of the factories of the big business groups like Samsung, Hyundae, Daewoo, KIA Motors and Lucky Goldstar.

Since this study is concerned with a critical comparative analysis of steel development in both Nigeria and South Korea in their post-colonial periods, its major analytical thrust is centered on the periods since the attainment of political independence in 1960 in the case of Nigeria, and for South Korea, in 1945. References are, however, made to the colonial and pre-colonial periods of the history of industrialization in both countries in order to provide a background context for a deeper understanding of the contemporary levels of growth in the steel sub-sectors of both Nigeria and South Korea. Since the state is at the heart of the steel development of both countries, the discussion is based mostly on the public steel companies. Reference is also made to the private steel companies inspite of the fact they account for very little in the activities of the steel sector. Lastly, the steel industry as used in this study, comprises of a group of both public and private companies that engaged in the production of iron ore, coal, limestone, and the alloying minerals as well as in the making of iron and production of steel products.

1.5 STRUCTURE OF THE STUDY

With this section serving as Chapter One, the rest of the study is structured into six chapters. Chapter Two sets the theoretical context of the study, and overcomes the basic unsettled theoretical issues emanating from the literature in order to create a more rigorous basis for understanding and comparing steel and industrial development in both Nigeria and South Korea. Chapter Three takes a comparative and critical look at the role of the state in the development of steel in both Nigeria and South Korea. An analysis of the structure and operations of the steel industries in both countries is undertaken in Chapter Four. Chapter Five critically evaluates the performance of the steel industries of both Nigeria and South Korea, paying particular attention to their product mix, technological route and interlinks with the capital goods-producing companies. Chapter Six examines the problems and prospects of the steel industries of Nigeria and South Korea. Chapter Seven, the Conclusion, suggests policy recommendations, stressing on the formulation of an appriopriate industrialization policy, and a special focus on what needs to be practically done especially by the Nigerian government in order to put the country's ailing steel sector back on course. All tables and figures are located in the Statistical Appendix.

NOTES AND REFERENCES

1. See, Daniel Omoweh, 'The Nigerian Steel Sector in the Global Steel Industry', *Annals, Journal of the Social Sciences of Nigeria*, no.8, 1996.
2. *ibid.*

Chapter Two

Contending Theoretical Issues

2.1 Introduction

This chapter first undertakes a brief but critical review of the literature on industrialization in Nigeria and South Korea. Thereafter, it identifies and clarifies basic unsettled theoretical issues arising from the literature in order to create a more rigorous basis for comparing the quest of both countries with industrialisation and steel development.

Until the formal colonization of Nigeria by the British in 1900, there were primordial cottage foundries and iron and steel works on which its process of industrialization had rested. The manufacturing of some basic farming implements like hoe and cutlasses, among other intermediate goods, was done by most of the pre-colonial cottage industries. In the period of British rule, the colonial state used its policies and actions to annihilate the cottage industries for the purpose of stalling all forms of competition which locally made intermediate goods would pose to the British-made ones that were imported into the colony. All was aimed at guaranteeing the Nigerian market for British export of intermediate and capital goods. And as a consequence, the locally made hoe, among other agricultural implements, with which the Nigerian peasant farmer had entered into colonialism, was replaced with an imported one from Britain at the time Nigeria attained its 'flag' independence in 1960. Today, the Nigerian peasant farmer depends on imported hoe and cutlasses, or those reproduced locally by foreign capital among other farming implements, as the major inputs for producing them are still imported. To break this jinx, the Nigerian state has to put in place, a vibrant iron and steel sector whose products would, in turn, be used for the production of both intermediate and capital goods.

Yet, since the Nigerian government established two integrated steel companies and three rolling mills in the country in the early 1980s, the situation has hardly improved as these companies have no production lines where the inputs for the production of capital goods are rolled out. All this has worsened Nigeria's technological dependence. The failure of the Nigerian state to get its steel sector really started much more talking of how it could be used to launch the country's industrial age, remains a major concern for scholars, policy makers, and all those who are poised to overturning its table of economic and industrial underdevelopment. This is all the moreso as other developing countries like South Korea, which has none of the basic steel raw materials compared with Nigeria, has recorded a relatively more impressive growth in its steel sector. In the light of Korea's 'relative success' with steel development, there is the general thinking among scholars including the author that , something should be fundamentally wrong with the Nigerian state and its approach to industrialization generally and steel development in particular. As noted earlier, this is not to assume that South Korea's impressive performance in its steel sector and industrialization at large, is not without some flaws. Let us review some of the major studies on industrialisation in Nigeria and South Korea.

2.2 A POLITICAL ECONOMY OF INDUSTRIALIZATION AND STEEL DEVELOPMENT

To begin with, the industrialization of both Nigeria and South Korea was not on the agenda of their colonizing imperial powers, Britain and Japan respectively. Suffice it to note that, it was only since the post-colonial periods of both countries that efforts were made by the state to promote some measure of industrialization. Since industrialization was not on the agenda of the colonial state, it did not bother to create a local business class that would undertake the task of industrialization following the termination of foreign rule. It is natural therefore, to expect the post-colonial state in both Nigeria and South Korea, to embark upon steel and industrial development with the hope of reversing their economic backwardness. What path did Nigeria and South Korea therefore, take to industrialization and steel development? Let us begin the discussion with Nigeria.

2.2.1 NIGERIA

Peter Kilby's study on Nigeria's industrialization remains one of the foremost that has been done so far. Concerned with what path he considered a relatively 'open' Nigerian economy ought to have taken to industrialize between 1945 and 1966, Kilby examined the structure of the country's economy, taking cognizance of the nature, capacity and role of the local productive forces, supporting institutions, labour, entrepreneurship and the professionals in Nigeria's quest for industrialisation. According to Kilby, since the end of the World War 11 in 1945, and even uptill 1960 when Nigeria gained political independence, the colonial state had, together with the forces of British imperialism, generally foisted import substitution strategy of industrialization on the country; that during the decade of the 1950s, the Regional Government Development Corporations of the North, West and East adopted the same import substitution industrial policy of the colonial state when they intervened into the economy of these regions; and that all this was done without any clear-cut industrial policy that would reverse Nigeria's economic underdevelopment. As a result fo this, Kilby continued, the cement, textile, tobacco, dairy and cosmetic companies among others established by both the central and regional governments were merely engaged in 'touch-up operations'. Worst still, the production lines of these companies were installed with refurbished machines most of which collasped either during trial production or packed up soon after they commenced real production. Beyond 1966, Kilby did not see the possibility of any industrial policy being formulated by the Nigerian state as its custodians were and still not interested in the country's industrialization. In resting his case, Kilby submitted that, until the basic economic and political fundamentals were put right, the chances of Nigeria taking advantage of its relatively 'open' economy to industrialize were slim.[1]

Nigeria's chequered history of attempts at industrialization since 1960 has shown that, afterall, Kilby was not an alarmist nor a pessimist. For, the Nigerian government had, at the time of this study, no industrial policy. Rather, what previous regimes in Nigeria had done since 1960, was to package incentives aimed at attracting foreign investors; incentives that have, rather than strengthen, weakened the country's industrial base as they help foreign capital consolidate its hold of the local economy. For instance, inspite of all the government's adherence to the import substitution strategy, what has taken place in practice, is the local reproduction of hitherto imported light and intermediate goods by the subsidiaries of the transnational corporations which also source their basic production inputs from their home countries. What is important to note, is not that the import subtitution strategy of indutrialization is impossible to achieve in Nigeria as South Korea's relative successful experience with that strategy showed that it could be realized given the right fundamentals. Its success in Nigeria therefore, is dependent on the state managers' political will and commitment to industrialization and steel development. However, if, unlike in South Korea, the subsidiaries of the Euro-American transnationals are still in control of the real sector of the Nigerian economy like manufacturing, then it will continue to impede its quest to industrialize. Adebayo Olukoshi's study sheds light on the kind of industrialization that the transnational corporations promoted in Kano, thereby bringing into fore the nature of the Nigerian state and its approach to industrialization.

According to Adebayo Olukoshi, though the subsidiaries of the transnationals promoted some level of industrialization in Kano, it was a form that did not really empower the local business class to be self-sustaining much more competing with foreign capital. Covering the period, 1903-1985, his study also demonstrated that, during the period of British colonial rule, the primordial cottage factories for which Kano was noted, were particularly badly destroyed by the policies and actions of the colonial state. And that in the early part of the post-colonial period, most of the subsidiaries of British transnationals and a fraction of Levantine capital had established factories that were engaged in the local reproduction of light consumer goods, while others were involved in the assembly of intermediate and capital goods. The decision by foreign capital to reproduce locally some of the light consumer goods hitherto imported, was in compliance with the former's policy to shift from purely exchange to 'touch-up operations' under the aegis of import substitution strategy. It was all an attempt by foreign capital to gain a firmer grip of the economy of Nigeria in the post-colonial period. Olukoshi's study also showed that, the nascent local capitalist class in Kano was, on account of its mercantile origin and weak production base, engaged minimally in the local reproduction of light consumer goods like candle, soap and perfume. Not only that, as some foreign companies still controlled the local assembly of intermediate and capital goods previously imported such as sewing machines, bicycles and their spares.[2]

In extending the frontier of his earlier study on industrialization in Kano to in the period since the Nigerian state adjusted in 1986, Olukoshi appraised the performance of the small capital-based manufacturing companies. His study showed that, under the adjustment situation, most of the manufacturing companies in Kano closed shops due largely to their difficulty of procuring foreign exchange with which to source for production inputs and spare parts of production machineries. All this resulted in increased cost of production. His study also showed that, the chances were remote for most of the companies which had closed shops since the onset of the adjustment to re-open them, thereby setting in motion, a trend of de-industrialization which has continued into the post-adjustment period. To Olukoshi, the trend could however, be reversed if the right micro and macro-economic policies and favourable political environment were put in place to enable the economy come back to keel.[3]

Without any doubt, Olukoshi's studies have provided us some useful insights into the nature and process of industrialization in Nigeria particularly his analysis of how the activities of subsidiaries of the transnationals had come to shape the structure of industrial development in Kano after having first annihilated the pre-colonial cottage factories in the colonial period. Sustainable industrialization would demand for a high local content of the production inputs among others, which the transnationals were not prepared to promote in Kano. Reversing the outward-oriented industrialization strategy which foreign capital had set in Kano, requires having iron and steel companies fitted with production lines that could roll out flat steels that are used to produce machines and their spares. Once this is done, integrated steel companies and other capital goods producing companies should be established to produce major machines and their parts. In essence, part of getting the basic fudamentals right as Olukoshi has suggested, should incorporate the regeneration of the country's iron and steel industry.

Though the nature and process of industrialization that has taken place in Nigeria since the turn of the 1970s, is still entrenching foreign capital into the real sector of the economy, the state has, more than ever, intervened significantly into some of the strategic areas too. However, the magnitude of problems that have accompanied the state's mode of intervention into the economy, have thrown up a lot of theoretical issues most of which were not treated in depth in earlier studies by Kilby and Olukoshi. Some of these issues for instance, centre around the model of industrial accumulation in Nigeria, nature of the policy framework that had informed the country's industrialization, and whose interest the state is really protecting. These were some of the issues tackled by Bright Ekuerhare in his most recent study on the pattern and problems of industrial accumulation in Nigeria.

According to Ekuerhare, the nature of industrialization in Nigeria particularly since the 1970s, has been state-led. So, too, is its model of industrial accumulation. This is all the more so since the Nigerian state is itself, capitalist. His study identified the absence of a well articulated industrial policy as one of the major problems that has deepened the country's industrial backwardness and constrained its own industrial accumulation generally. To Ekuerhare, the chances of overcoming these problems were remote since state capitalism is still basically commercial. As part of its commercial capitalism, his study showed that, the state always sought alliance, albeit an unequal one, with foreign capital, as it intervened into the economy. The country's worsening technological dependence on the transnationals was identified as one of the major consequences of the state model of industrial accumulation. He cited the indiscriminate importation of production machineries and spares to instal textile factories without developing the local technical capacity to reproduce them to shed light on the nature of the country's industrial and technological backwardness. To enable the state transcend its predatory mode of accumulation, Ekuerhare advocated a socialist state that would be oriented towards self-sustaining economic development.[4]

That the Nigerian state is capitalist, and the same for its model of accumulation as Ekuerhare has pointed out, is well taken. Without any doubt, locating his analysis of industrial accumulation within the framework of the dialetics of history has provided us a deeper insight into the nature of Nigeria's industrial backwardness. However, given the failed attempts of some socialist governments in Africa like then Nyerere's regime in Tanzania, the Mengistu's government in Ethiopia to bringing about economic development of their respective countries, then a socialist state that Ekuerhare has suggested for Nigeria, might not be the real anathema to the country's industrial backwardness. It is also not clear in his study, how the proposed socialist Nigerian state would emerge. For, the issue at stake is not that those in charge of the state are unaware of their collusion with foreign capital for the parochial gains and its implications for the economy, but how to make them rethink their approaches to industrialization. That is not all; for, having a socialist state does not really presume that the right fundamentals will be got right since if it were so, the collapse of the then Union of Soviet Socialist Republic [USSR] would have been avoided. In fact, One agrees with the position of the Nordiska African Institute, Upsalla-sponsored conference on what path Africa would take into the 21st century, in which it was cautioned that African governments and scholars should exercise caution in prescribing socialist policy as a way out of the continent's quagmire. For one thing, past experiences have shown that, hitherto, socialist states in Africa have abandoned Marxist policies to embrace the market reform of the Bank and the Fund. Furthermore, the economic crisis that has plagued former socialist countries in Eastern Europe is, in part, an indication that the panacea to the deterioration of the economies of African countries does not basically reside with ideology. While a deep understanding of the state model of industrial accumulation as Ekuerhare did in his study, has provided us important insights into the nature of the country's industrial backwardness, the capital goods industry was treated in passing; yet, it constitutes the repository of the local capacity that would bring about any meaningful technological breakthrough. For instance, the South Korean government was partly able to overcome basic aspects of the limitations of import substitution strategy after its steel companies had started manufacturing production machineries and their spares hitherto imported.

Until the Nigerian government established two integrated steel companies and three inland steel rolling mills between 1982 and 1983, activities in the country's steel sub-sector was not only in the hands of small private steel companies, but really insignificant. As a result, there were basically no serious scholarly studies conducted on the Nigerian iron and steel industry. However, the few studies that were udertaken on the country's steel sector came after the state had adjusted in 1986. Even then, most of the studies were still no less than an appraisal of the public steel companies performance under the adjustment period, while the rest examined the country's steel sector outside of the context of adjustment. Let us begin with the latter set of studies.

In his contribution to the CODESRIA-sponsored study on accumulation , capital goods and technological change in Nigeria, B.U.N Igwe argued that politics had, rather than techno-economic considerations, informed the establishment of the state-owned iron and steel companies. And as a consequence, they were not fitted with production lines that roll out flat steels, I-Beams, and capital goods that would be used in endogenizing technology. Rather, the steel companies produce only simple constructional steel products like coils and rods.[5] Closely related to Igwe's argument is Bayo Kolade's, whose analysis of the operations of the Oshogbo Steel Rolling Mill showed that, the company was not in anyway linked with the capital goods industry. Why? According to Kolade, it was in part due to the fact the company as an end-product operation, rolls out basically that steel rods and bars which are rarely used in the capital goods industry.[6] To reverse the trend, both Igwe and Kolade called for the immediate formulation of a steel policy that would guide development of steel and industrialization in Nigeria.

Beyond identifying these problems and their call for a steel policy to be formulated, it is not too clear in the studies by both Igwe and Kolade what perspective the proposed steel policy should take: Is it to be state-led or driven by private capital; or what? Furthermore, there was really no concrete suggestion on what should be done to regenerate the ailing public steel companies which have been bedevilled with crisis from their inception. For instance, it is possible for the rolling mills to be integrated backwards by equipping them with the facilities that will enable them consume scrap metals or even iron ore and produce flat steel sheets as Kolade has suggested. But he did not suggest ways to get the state re-organised in such a way that it will be really interested in steel development beyond its current cosmetic efforts. Without settling this political issue, it can be argued that the country's quest for the endogenization of technology in particular, and industrialisation generally, would remain elusive. Now let us treat other studies that have been conducted on the Nigerian steel sector since the period of adjustment.

Although virtually all other sub-sectors of the Nigerian industry were faced with problems of shut-down, closure of shops, down-sizing and negative capacity utilization since the Nigerian state embraced the World Bank/ IMF-sponsored adjustment programme, the country's public steel companies were particularly badly hit. Taking off without an industrial /steel policy, and at the period when the global steel industry was itself, in crisis, the public steel companies which account for about 97 percent of the activity in that sub-sector, were not founded on viable and self-sustaining techno-economic grounds. As a result, barely a year after the steel companies had taken off, the government could hardly finance them as evidenced in the acute shortage of raw materials, spares for the production plants resulting in the cannibalisation of machines, frequent closures and redundancy of staff. All this was to help frame the studies that were later undertaken on the country's iron and steel sector.

In 1988, the Bank and the Fund had, in concert with the Paris and London Clubs to which Nigeria was heavily indebted, and the Western European steel transnationals that built the country's public steel companies, commissioned the Hatch Associates of Canada to appraise the ailing state-owned steel companies and recommend a plan of action for the sponsors. According to Hatch Associate's report, the public steel companies were ridden with crisis which included over-investment, poor management and unduly large size. The highpoints of its recommendation were: i] that the Ajaokuta steel project in which about US$3billion was spent by 1986, was yet to be completed, and therefore, suggested that it should be abandoned. Another reason suggested for its abandonment, was Ajaokuta's installed capacity for crude steel, which was considered unduly large in comparison with country's steel need; ii] that the Delta Steel Company should be leased out on contract to foreign capital; iii] that the steel rolling companies should be privatized; and iv] that the National Iron Ore Mining Company at Itakpe, should produce iron ore basically for export.[7]

The recommendations of the Hatch Associates which the Bank/Fund, in turn, presented to the Babangida's administration as conditionalities for loan to regenerate the public steel companies were turned down. Why? The custodians of the state saw it as plot to fully retrench the state from the country's steel sector in order for the Western European steel transnationals to take it over. In other words, the real intention of the World Bank and the IMF was not to help the Nigerian government rehabilitate its ailing public steel companies, but to help protect the country's huge steel market for foreign capital. Against this background, then regime of General Babangida established the Technical Committee on Privatization and Commercialization [TCPC] which was later renamed the Bureau of Public Enterprise [BPE] to work out alternative modalities for revamping the country's crisis-ridden steel companies among other industrial sectors of its economy. Before dwelling more on the TCPC, it is important to note that, as of March 1999, the junta regime of General Abdulsalami Abubakar sacked the BPE and instituted a Council on Commercialisation and Privatisation under the chairmanship of Admiral Mike Akhigbe, Chief of General Staff.

The TCPC commissioned Dafinone and Company, a firm of chartered accountants based in Lagos, Nigeria, to carry out an evaluation study of the Delta Steel Company and the three steel rolling mills with a view to suggesting ways to regenerate them with particular emphasis on funding. According to Dafinone and Company's report, though on paper, the steel companies were established as joint stock ventures, they operated as government parastatals in practice. This was identified as a major contradiction inherent in the approach the state took to steel development; a contradiction that partly accounted for the problems of over-investment, liquidity squeeze and poor management that the steel companies were faced with. In its recommendation, the company suggested that DSC should be commercialized as a measure of reducing its heavy financial burden on government. And for the steel rolling companies, the company recommended that they should be privatized moreso as they were end-product operations.[8]

As a creation of the Nigerian government, the TCPC accepted the recommendations of Dafinone and Company, though with major modifications. With respect to DSC, the TCPC was of the view that its commercialization should, rather than being guided by the forces of the market, proceed under the aegis of the state, with the majority of its equity position still to be held by the state. The same was true of the privatization of the steel rolling mills in which the state also should be the majority shareholder. There were other studies by some independent consultants such as the one which Katsina Steel Rolling Company commissioned the author to undertake in 1992, and in which he recommended that the state should privatize the company by selling substantial shares of the company to steel workers and other members of the Nigerian public. By doing so, his Committee was of the belief that it would help create a viable local business class that would complement the effort of the state in developing the steel sector. As to be expected, the government did not accept its recommendations.[9] Suffice it to note that, be it commercialization or privatization, the state is clearly unwilling to let go the public steel companies. Why? It is because the steel companies, among other big state-owned companies, are used to lubricate the state model of extracting surplus from the economy, and giving plum but highly unproductive appointments to its cronies.

Like the previous studies on Nigeria's steel sector since the advent of adjustment, the contributors to the 12th Annual Conference of the Nigerian Metallurgical Society [NMS] held in November 1994, all agreed that the basic problem of the steel companies was finance. And they contended that, what was needed to turn the ailing steel companies around, was to fund them adequately.[10] One of the main reasons why liquidity freeze was identified as the most serious problem of the public steel companies stemmed from the fact that the major sponsors of the conference were themselves, managers of the public steel companies and therefore, took the opportunity of the meeting to publicly register their request to the government. There is no doubt that the steel companies were actually mired in financial crisis and that more funds ought to be injected into them if they are to recover. However, one of the fundamental unresolved issues in the development of the country's steel sector is not so much with funding as to whether iron and steel development is really on the agenda of those entrusted with the management of the Nigerian state. As the author has argued elsewhere, the Nigerian state has, in an effort to further its accummulative base, embarked on the development of iron and steel by recycling part of the oil rent to start off other rent-yielding concerns. In essence, the steel companies were not really set up to launch the country's industrial age. It is not that the state managers were unaware of the dire consequences of the approach they took to steel development for the country's economy and industrialisation. Rather, the ultimate aim of the state managers is to lease out the steel companies to both local and foreign private capitals to manage and for which they will pay rent as already evident in the current mode of managing the completed public steel companies. All this was to structure the nature of state's funding of the steel companies and how they are managed.[11] Suffice it to say that, the current poor status of the steel companies is indeed, what the state wants them to be. This contrasts sharply with the path which the Korean state took to the development of its iron and steel sector.

2.2.2 SOUTH KOREA

In order to have a deeper insight into the path which the Korean state took to steel development, let us first have an overview of economic development in South Korea since 1945. South Korea's relatively impressive growth in its steel sector in the last 25 years particularly in comparison with Nigeria's unsuccessful attempts at steel development, has endeared the former's industrialization policy and projects to policy makers, government officials, scholars and agents of development world-wide. In fact, the 'rise' of the Korean economy in the 1980s, a decade considered to have been lost in Africa including Nigeria, was understood by many a scholar, as a celebration of both the neo-liberal economic theory and the market. In the theoretical discourse on the newly industrialising economies in East Asia, one of the major reasons that was identified for the dramatic industrial and economic growth of these countries not the least South Korea, was the ability of its policy makers to have got its 'basic fundamentals' right. The World Bank's study on the East Asian NICs in 1993 did not only endorse the development policies of these countries, but also, recommended them for adoption by the governments of African countries whose economies were still faced with severe political , social and economic crises.[12] That is not all. South Korea's impressive economic growth has marvelled Koreans themselves, forcing the majority of them to believe that it is not unconnected with the spiritual power of the 'Huan River'.[13]

This study is not refuting the fact that significant industrial and economic growth was recorded in the Korean economy, particularly in comparison with Nigeria and other African countries. For, it is obvious, and even its relative success as a 'late industrializer' helped raise the hope of other developing economies like Nigeria that, it, too, could become developed both industrially and economically someday if all the right fundamentals are got. In fact, Korea's impressive economic growth is a demonstration, that though Japanese colonialism had underdeveloped the peninsula in the past, with a committed, strong and purposeful Korean state, the country's economic and industrial backwardness was largely reversed. Rather, my contention here is that, in spite of Korea's rapid economic growth, there are many unsettled theoretical issues in the approach the Korean state took to industrialize, which must be resolved in order to deepen the understanding of the nature of its micro and macro-economic policies that informed its development strategy, and then, undertake a rigorous comparison of its economy with other developing countries such as Nigeria. For instance, while it might be true that the Korean government got its 'fundamentals right', it is equally important to understand the political regime both locally and globally, under which its policies were formulated while comparing South Korea and other Asian NICs with African countries. All this is done in an attempt to determine whether the policy regime that permitted South Korea's rapid industrial growth inclusive of its steel sector generally, can indeed, be replicated or not, in Nigeria among other African countries.

Since the turn of the 1980s, there has been a growing tendency among scholars of Afro-Asian studies to want to compare the Asian NICs such as South Korea, Taiwan and Singapore with African countries like Nigeria and Ghana. At issue, is the foregone conclusion with which such comparisons were done. Not surprisingly, the Asian NICs were seen as models for Africa whose economy is plagued with crisis and with little or no prospect of recovery in sight. Some scholars are still of the view that, there is indeed, no basis for such comparison given the rapid economic growth of the Asian NICs vis-a-vis the deteriorating economies of Africa. What is too easily forgotten in all these discussions, is that, in the 1960s, most of the fast growing economies in Asia were not significantly different from their counterparts in Africa in terms of the shares of agriculture, manufacturing, mining among other economic indicators, in the Gross Domestic Product.[14] Not only that. Scholars rely heavily on some of these basic economic indicators such as *per capita* income, growth rate and gross national product in trying to bring out the distinctions in the economies of the countries in the two regions, indicators whose underpinning factors are not clarified, could obscure the details of the reality on ground, and should not really be a rigorous basis for recommending Asia's approach to economic development for Africa. For instance, Deborah Brautigam had, in comparing Taiwan with Sub-Saharan Africa, noted that while the former recorded a national economic growth rate of 8.2 percent, the latter grew at 0.5 percentage point within the 1980s.[15] Taiwan's sustained growth in the thinking of Brautigam, should serve as a lesson for Africa. As a 'late industrializer', Taiwan's impressive rate of economic growth with some measure of equity should be a source of inspiration for Africa.

However, with a deeper understanding of the political economy of rapid industrial growth of the Asian NICs, it will be misleading to make such a claim as Brautigam did. For, with such background knowledge, it will become clearer that a lot of unsettled theoretical issues will have to be clarified to avoid such mistake. So, while these basic economic indicators might be useful basis for comparing industrialization in Africa and East Asia, there is the need to contextualize their analysis, while paying serious attention to the nature of the state, its production systems and how all this, in turn, influenced the overall structure of the economy. These factors, no doubt, not only differ from one region to the other, but also, significantly determine the kind of data that are generated. As noted, it is the clarification of these basic unsettled theoretical issues arising from the review of studies on Nigeria and South Korea that will then lay the basis for comparing their experiences with industrialization generally and steel development in particular.

To begin with, for 35 years, 1910-1945, that the Japanese colonized the Korean peninsula, the policies and actions of the colonial state did not permit its industrialisation. It was a delibrate policy of the Japanese colonial state not to create a local business class capable of undertaking the task of promoting industrialization since it had no plan to let Korea become politically independent. The fact that the Japanese never thought of an independent Korea partly accounted for the nature of the industrial underdevelopment of South Korea in 1945 when it finally became a sovereign state. Even in Nigeria where the British reluctantly granted it political independence without the use of force, its industrialisation was also not on the agenda. So, like its counterpart in Nigeria, it is natural to expect the postcolonial Korean state to embark upon the country's industrial development particularly if it wanted to reverse the country's economic backwardness at independence. However, in the next 15 years after Korea's independence, the country's industrialization was not really on the agenda of those in charge of the Korean state, a situation that was similar to Nigeria's. Why was it so? One of the major reasons was that, the United States of America (USA) government which helped liberate Korea from the Japanese colonial rule, stayed back in the country to protect its strategic and political interests in the peninsula. That is the origin of how the US used South Korea as a 'bulwark' to check the expansion of Sino-Russian communism into the South East Asia/Pacific region during the cold war period. There was no geopolitical calculation by the superpowers in Nigeria during the same period. Under such political circumstance, in addition to Korea's geographical proximity to its neighbour in the North, the US foreign policy towards South Korea shaped its internal political and economic development. For instance, South Korea's border with North Korea made it vulnerable to military incursion by the latter, which overran the former during the three-year war, 1950-1953. This compelled the custodians of the Korean state to seek for a security cover from a more powerful country like the

US. And as a result, the Korean state could not stand off against the geo-strategic interest of the American government. One of the implications of all this, is that, while the Korean state lived off the goodwill and benefitted from the flow of aids from the American government, it was particularly handicapped from embarking on any serious economic development of the peninsula. It is all the moreso since South Korea was poor both in terms of finance and natural resources with which it could embark upon the economic and industrial development of the country.

It was not surprising therefore, that by 1960, that is, 15 years after South Korea became a sovereign state, and the year Nigeria gained political independence, the Korean economy was still in the woods and had similar level of basic economic indicators with the majority of African countries. For instance, South Korea's *per capita* income of about US$95.00 in 1960 was almost equal to Nigeria's US$93.00. By 1960, too, South Korea was still faced with deepening poverty and overall worsening social, political and economic crises so much so that the majority of Koreans had lost hope on the capacity of the Korean economy to ever recover. The pessimism stemmed largely from the colossal destruction of lives and properties during the 1950-1953 Korean war. So, too, were the Western European countries pessimistic about the prospects of economic recovery of the peninsula from the wreckages of the three-year war. The US government's interest in geostrategy and not the economic development of the country further worsened the prospects of recovery of the Korean economy after the war.

Why was South Korea a write-off in the 1960s? As noted earlier, in using South Korea to stem the incursion of Sino-Russian communism in the South East Asian region, the United State Military Government in South Korea first took charge of the administration of the penninsula for three years, 1945-1948, before it installed the Rhee's government in 1948. Sygman Rhee, it should be noted, was not only invited from the USA where he was a university teacher, but was the consensus choice of the American government for the Korean president. So, between 1948 and 1961 when the Rhee's government was in power, there was really no plan to have the industrial development of Korea on the agenda. As a result, the country's economy had, rather than improve, remained in more severe crises: acute shortage of food deepened, lack of basic social amenities worsened, the productive base furthter weakened, and the illiteracy level hightened.[16] Given the nature of the American government's interest in the Republic of Korea, it was not surprising that the Rhee's regime could not really do much to bail out the aid-dependent Korean economy from these crises. As the socio-economic problems which the Korean economy was faced with bit harder, it heightened the fears of annexing the Republic by North Korea. It became obvious that one of the immediate ways that could lead the country out of its deepening political, economic and social crises, was to change the government in power. And the change of guard was done undemocratically through a military coup in 1961.

So, in 1961, when General Park Chung-hee and other Korean army officers seized political power in a *coup d'etat*, a new sigh of relief was heaved by Koreans across the country. The Park regime was firmly convinced that, the nature of the real crisis which the Korean economy was faced with then, was economic and not political even though its root cause was political. Therefore, the General Park administration was of the view that, for South Korea to overcome its industrial and economic backwardness, the country should stop its 'beggar thy neighbour policy', and begin to embark on its own economic development. Because Korea lacked basic industrial raw materials, had no strong local business class that would engage in real production of goods and services at independence, and poor financially, it was natural to expect the state to lead the country's quest for industrialization. In the light of this constraint, coupled with the relatively more developed defence industry of North Korea, the Park regime placed high premium on Korea's industrialisation. Apart from the security reasons earlier mentioned, other factors that prompted the Park's regime to begin the industrial development of the Republic emanated from the outside. The initial unwillingness of Japan and the majority of Western European countries including USA to advance the Park administration loans with which to embark on the country's economic and industrial development inclusive of iron and steel, made it all the more determined to get Korea's industrialisation underway. One of the hallmarks of the economic policies of the Park's regime, was its strong emphasis on growth, while it paid less attention to other equally serious development issues like equity and stability. Given the hard times that the Korean economy was faced with, the primary concern of the Park government was how to make the economy grow; thereafter, issues of equity and stability would be settled. This contrasted sharply with the development startegies of Taiwan and Singapore where relative balance was struck between growth, equity and stability. Not only was the Park government so growth-obsessed, it placed the state at the heart of the economy and the institutions established to implement its industrialization policies and projects.[17]

With the military regime of General Park firmly in control of the Republic of Korea, its development strategy particularly for industrialization could not have proceeded without an initial reparation of $300million paid by the Japanese government to the Park government in the 1960s. This elicited a change in the hardline policy of the West towards South Korea, a change that was further made possible by the politics of the cold war in the Asian sub-region. The West began to give loans to South Korea largely to contain the influence of China and then USSR in the region. From external funds, the Park regime was able to begin a state-led economic development programme in Korea in the early 1960s. By the 1970s, the economy of the Republic had already begun to grow so much so that, in the next decade, it recorded 8.5 percent. Concerned with its growth policy, the Korean state got itself heavily entrenched into the heart of the country's economy, while it created the big business groups that lubricated its mode of surplus extraction from the economy. The big trading companies or *chaebols* were, indeed, created by the Park government ostensibly to implement the export-led industrialization programmes. In fact, they were fused with government that it was difficult to draw any distinction between them. That is not all, as in the financial sector, the state was also in charge and in most cases, acted as the sole guarantor of investment loans that were advanced to the big business groups. The same was true of the heavy industrial sector of which the steel sub-sector was a major part and its prime mover. From all indications, the Korean state looked very committed to Korea's industrialisation. Some of the evidences of this commitment was: the Republic's annual economic growth which rose from 1.5 percent in the mid-1960s to 8.5 percentage point in 1980s. With an

8.5 percent annual growth rate in the 1980s, the Korean economy represented one of the fastest growing economies not only in East Asia, but among the developing economies since the end of the World War II in 1945. What is more, without iron ore, coal and alloying minerals, but an autocratic state that was committed to industrialisation inclusive of steel, South Korea became the sixth world largest producer of crude steel accounting for 5.32% of world's total of 730,494,000 tonnes in 1996. South Korea also ranked as the largest exporter of microchips in 1996, and second largest ship-building country in the world after Japan in the past five years.[18]

Korea's industrial feat in the past 25 years inspite of all the odds placed on its path in the past, had endeared it to policy makers and scholars that it started attracting tremendous scholarly studies and in a magnitude that was never recorded since the country's independence in 1945. Having provided the background context, it is in order to return to the discussion proper.

Studies on Korea's industrialisation could be collapsed into two broad groups going by the methodologies adopted by the scholars. First, is a group of scholars whose major methodological approach is largely a description of the structures of the economies of the Asian NICs inclusive of South Korea. In this group, the focus of study was basically centered on the development policies, strategies and institutions of the Korean government, noting the role played by the big business groups in bringing about the industrial growth of the Republic. Notable scholars in this category, included Alice Amsden[19], Byung-Nak Song[20] and Tae-Wan-Son.[21] Let us first discuss these studies before proceeding to examine the second group of scholarly writings.

In her pioneer study of the Korean economy since the 1960s, Alice Amsden gave the most comprehensive account so far, of how the South Korean economy rose from 'dust' in the early 1960s to attain an enviable economic status in the 1980s with great prospects of becoming the Asia's next industrial giant after Japan. Tracing the history of Korea's economic growth, her study demonstrated that, what could ordinarily be regarded as the 'secrets' or 'miracles' behind the relative success of Korea's 'late industrialization', was no less than the authoritarian-bureaucratic Korean state and its development strategy; a strategy that only began in 1961 when General Park overthrew the regime of Rhee. Amsden's work pointed out that the strategy which informed Korea's economic and industrial development was rooted in growth, though at the expense of equality and stability. Under the development strategy, she continued, the state set the targets' goals and formulated the general industrial policy framework to accomplish them, and within which Korea's late industrialization proceeded. Also, Amsden brought out very clearly in her study that, together, the readily cheap and committed but fairly skilled Korean labour generally, and the cadre of dedicated salaried engineers in particular. who acted as 'gatekeepers' of foreign 'technology transfer', contributed tremendously to the country's industrial rapid growth. Drawing on the particular significant progress recorded in the Korean automobile and steel industrial sub-sectors, Amsden's study showed that, the country had recorded a leading position in technological advancement in South East Asia. In her submission, Amsden encouraged other developing countries to master the rudiments of Korea's approach to industrialization if they are to succeed as 'late industrializers'.[22]

Byung-Nak Song's study was not significantly different from Alice Amsden's in the sense that he, too, covered basically the same issues that underpinned the remarkable growth which the Korean economy had recorded, namely, the Korean state including its development policy and strategy, the role of the big business groups, the Korean people and their culture particularly the new Confucian ethic.[23] Therefore, investing time and space discussing Byung-Nak's study is not really necessary at this point. Rather, what is important to note, which is being done here, is to examine the context in which he raised these issues in his study. While Amsden's study had traced the strategy that informed the 'relative success' of South Korea's late industrialization to the Korean state and did not hestitate to recommend it, though with some caution for developing economies interested in mastering it, Song's major contention was that, it was, indeed, 'Korea's model of development' in spite of all evidences that it was nothing of the sort.

To Song, what he called the Korean model was, indeed, really the brainchild of a crop of Korean intellectuals of which he is one. Song saw the so-called Korean model as a reference for other Third World countries that are really interested in getting industrialised particularly in the 21st century. Departing from Amsden, and in an attempt to shed more light on the ingenuity of Korean policy makers and scholars especially the role they have been playing since the advent of the Park administration to help think out the policy framework for Korea's rapid economic and industrial growth, Song compared South Korea with Japan, another late industrializer. His comparative analysis of both Korea and Japan was based on indices such as growth pattern, trade and incentive system, industrial structure and policy, economic planning and policy formulation, equity of income and economic life. To him, the Korean development model was homegrown, and that though the Korean economy shared some common features with that of Japan, it was not strictly patterned along the latter's approach to economic development as some scholars have earlier argued. Song was quite optimistic in his study that, sooner than later, the Korean economy, which was seen as having already attained 'late growth', would reach 'market maturity'; a stage that the Japanese economy was for almost two decades ago.

No doubt, both studies by Amsden and Song have provided us useful insights into the rise of the Korean economy particularly the character of the Korean state, its mode of accumulation, and role in the country's industrialization. While Amsden's recommendation that other developing economies willing to industrialize at the brink of the 21st century should be cautious in mastering the rudiments of Korea's approach to industrialization, it is not too clear what she meant by that. Could she be unwittingly endorsing an authoritarian state for the developing economies given the character of the Korean state since the advent of the General Park regime and subsequent regimes including that of the incumbent, Kim Dae-Jung and his successors? It is not in doubt that the Korean economy witnessed dramatic industrial growth in the period since the 1970s till the early part of the 1990s. At issue rather, is whether Korea's economy ever underwent any real industrial development, if it is taken as a process. For, industrial development, like any type of development, is a process that strives to strike a balance between growth, equity and stability. Even if it starts as the project of a particular class within the state structure as was the case in South Korea, it still has to address these three critical issues. But that is not the experience of Korea's industrialisation. Compared with Kilby's study of Nigeria, where attempts at industrialization proceeded without any an industrial policy guide, findings from Amsden's study also demonstrated the role that an industrial policy could play in bringing about the economic development of a country such as South Korea. As noted earlier, many factors actually went into the shaping of the industrial policy of South Korea, notable among which were, the nature of the state, the global political environment, and resource endowment. Yet, Amsden failed in her analysis to show how some of these factors influenced Korea's quest for industrialisation, and what policy lessons it all portends for other developing economies like Nigeria. The fact that, the historical past of these countries was never the same made it all more so.

As for Song, it is not in doubt that the Korean economy grew dramatically in the past two decades. However, what is totally missing in his analysis is the basis on which he recommended the Korean approach to industrial development as a model for other developing economies. The major question that arises is: How would the Korean industrial sector that is still at the stage of 'late growth' be a model?. While one partly subscribes to the position of the Korean government officials like those of the Korean Development Institute that, a country's leaders' choice of the path to industrial development matters a lot, it should not preclude the views of the governed as industrialization ought to be both a process and a collective enterprise of all. But from all indications, that was not the case with South Korea since it was no less than the project of a tiny cabal that constituted the General Park regime, a trend that was continued with after his overthrow in 1979. Furthermore, it did not really matter whether South Korea's approach to industrialization was patterned along the line of Japan. Rather, Song's study did not really develop any alternative philosophical framework to draw his distinction between the economies of both South Korea and Japan other than counterfactualizing. The emerging structure of the economies of the Asian NICs since the turn of the 1970s showed that, although they had devised different strategies to actualize their quests for industrialization, virtually all of them including South Korea, took after Japan's approach as a reference at one piont or the other .

Insightful as the studies by Amsden and Byung-Nak have been, they have left more unsettled theoretical issues in the literature on the Korean economy than addressed. At issue for instance, is not economic growth *per se*; rather, it is the nature of the policy regime in which it took place, its relationship with equity and stability, and the nature of the agents of economic growth like the big business groups. It is in an attempt to provide answers to these unresolved theoretical issues in the literature so that a rigorous basis can be created for comparing industrialization in the Asian NICs and Africa, that gave rise to the second group of scholars.

The basic contention of the second school of thought is that, the development strategy which informed South Korea's rapid economic growth and industrialization including those of other Asian NICs like Taiwan, Hong-Kong and Singapore, can hardly be replicated in African and Latin American countries. As for South Korea, the major contention was that, the orientation of its industrial policy was tilted towards the selfish interests of the state managers and away from the dominated; that its economy was closed and not driven by the forces of the market; and finally, that its peculiar historical experience during the Cold War could hardly be replicated in any part of the world.

Compared with African and Latin American countries, the second group of scholars contended that, though the decades of the 1970s and 1980s were indeed, lost in these regions, the autocratic regimes and non-compliance with the tenets of the market under which the Asian NICs recorded impressive economic growth during the same periods do not make them models for other developing economies. In essence, despite their remarkable economic growth, the Asian NICs were and are still isolated from the flourished theoretical debates on the political economy of development in Latin America and Africa. So, the major concern of these scholars notably, Fredric Deyo[24], Chalmers Johnson[25], Peter Evans[26], Ray Kiely[27] and Jong-Chan Rhee[28] and to which the author subscribes, is to first establish the theoretical basis that will help integrate the East Asian NICs' experiences with industrialization into the mainstream of the theoretical discourse on the social, economic and political development of Africa among other developing economies.

Therefore, the task of the next section is not to really embark on a review of the writings of these scholars, but to identify the major unresolved theoretical issues raised in the political economy of Nigeria and South Korea, with a view to clarifying them. All is in an attempt to create a more credible basis for undertaking a comparative analysis of the experiences of both Nigeria and South Korea with steel development and industrialization.

2.2.3 Overcoming Basic Unsettled Theoretical Issues Underpinning Comparing Industrialization in Nigeria and South Korea

The State: One of the basic issues that has clearly emerged from the above literature is that, both Nigeria and South Korea were once under colonial rule, and emerged from foreign domination not only with a weak state, but also, a local business class that is only interested in exchange and not production. Therefore, it is natural to expect the post-colonial state in both Nigeria and South Korea to intervene into the development process of the economy of these countries if the industrial underdevelopment they had inherited from the colonialism is to be reversed in the period of political independence. However, one of the distinctions between the two countries can be located in the role the state has played so far, in the process of industrialization inclusive of steel development. Yet, quite often, scholars pay little or no attention to this issue while undertaking comparative analysis of African and Asian economies.

One of the major hidden factors behind South Korea's rapid economic rapid economic growth among other Asian NICs, is the state and its role in the development process of the country. As late industrialisers, modernization theoretists like Samuel Huntington, has often advocated that, given the unwillingness of the developed industrial economies to really help the governments of Third World countries promote economic development, has necessitated the need for a 'strong' state that would reverse the underdevelopment of these countries. Little wonder, scholars like Huntington had advocated for other developing countries, an authoritarian state in the like of the Asian NICs, if they wanted to get their economies back on the right footing. To an extent, Huntington's thesis of the state yielded some positive result as evidenced in the rapid economic growth of South Korea and other Asian NICs like Taiwan and Singapore, where an autocratic state also led to the process of economic development. The question that arises from Huntington's advocacy of hard state is: Should Nigeria adopt an authoritarian state like the South Korean if it must witness economic growth? The Korean experience has shown that, under an autocratic political rule, economic development inclusive of industrialization could not be a collective enterprise as labour and the people hardly had any input into the decision-making system. Rather, it became the project of a tiny class within the Korean state structure, whose members, on account of their parochial interests, imposed on the people and the country, the path that Korea's industrialisation and economic development took. The implications of the undemocratic approach with which the Korean state had intervened into the economy began to manifest in the late 1970s when the Park regime could not embark on any meaningful industrial adjustment to redress the hyper-inflationary trend the country was faced with then, the recent Korea's economic crisis being the second signal of the unviability of Huntington's authoritarian state project. In fact, Huntington had created the erroneous impression that the hard state he anticipated in the developing economies would be class neutral and guide society rather impartially. It is for these and other related reasons that the author, together with scholars like Chalmers Johnson, Frederic Deyo and Peter Evans are opposed to Huntington's thesis on the 'hard state'.

Drawing on the experiences of Latin American countries with industrialization, Johnson and other scholars had argued that, inspite of the hard state of the countries in the sub-region, economic development policies and projects that were embarked upon did little to improve the well-being of the people and the economy at large. This was all the more so because the post-colonial state had its various institutional groups immersed in intra-class struggles for power and accummulation, and manipulated by foreign capital. In extending the analysis of the shortcomings of the hard state to the Asian NICs of which South Korea is one, the major contention of these scholars is that, the post-colonial Korean state, like its counterpart in Latin America and Africa, was partial. Suffice to note that, though state-led industrialization in South Korea promoted its rapid economic growth, it not only benefitted its groups' interests particularly the cabal in the Korean military, and owners of the *chaebols* more than an average Korean. Herein lies how the root cause of the fragility of the Korean economy can be traced to its authoritarian political regime. Extending the analysis further to Nigeria, it can be argued that, though the repressive Nigerian state had led the country's industrialization like its Korean counterpart, the Korean state was a relatively better performer. For instance, while state authoritarianism was able to bring about some measure of economic and industrial growth in Korea, the opposite was the case in Nigeria. Why? It is partly because there is more to the capacity of a state and its commitment to economic growth than just being a hard state. In other words, it is not enough to draw comparison between Nigeria and South Korea largely on the grounds that the state in both countries is autocratic. Rather, there is need to understand the character of the state and its mode of surplus extraction which provide more useful insights into its performance in the development process.

South Korea, like other Asian NICs of Taiwan and Singapore, had what one has chosen to call a 'hard state' in the sense that it was authoritarian with the military constituting its hegemonic class. The Korean state was an all powerful state and present in every aspects of the Korea's political economy. The Korean government, its executive organ, is in the words of Parvez Hasan, 'the senior partner and major participant who determined all that went on in the Korean economy'[29]. In fact, it was the policies and actions of the Korean state that structured the economy. Hence, its economy was still state-led, non-marketized, and closed inspite of all pretentions that it was the opposite. All has come to explain why it is most likely that no matter the rapid growth the Korean economy will record, as long as the state structure and the polity remain undemocratic, the gains arising from such growth, will sooner than later, be consumed by the same contradictions the system engenders.

In the light of the above analysis, it is difficult to fathom how apologists of the Asian NICs like Song characterized Korea's economic growth as a model, which he recommended for other developing economies like Nigeria to follow.

As noted, the Nigerian state, like its Korean counterpart, is authoritarian and repressive only in the sense that the military has captured political power for 29 out of the 39 years of the country's political independence as of 1999. In terms of an instrument of economic development, the Nigerian state was largely used for repression and vindictive politics. Unlike in Korea where the autocratic state was well established, committed towards economic growth, and in firm control of the economy, in Nigeria, there is really no state, whether liberal or autocratic. Instead, what exists in the form of a state is nothing but a public force and it is only nominal since it is basically privatized. Because of its privatisation, the state is unable to rise above the conflicts and struggles of its various constituent groups for the appriopriation of resources including the power of the state. In essence, the state is itself, a major platform of intense struggles between and among the institutions that make it a reality. In fact, it is the intensity of the struggles to capture state power and retain it by all means and at all cost, that have reduced politics to a warfare and zero-sum, all of which is to the detriment of development including industrialization. Under such circumstances, industrialization could hardly be on the agenda of the state managers as their ultimate interest is to capture political power by all means since it holds the key to wealth. Therefore, rather than change, those in charge of the state only merely inherited structures of colonial policies of exploitation unlike their counterparts in Korea, who under an autocratic regime, closed the country's economy while grappling with both the internal and external challenges of economic growth. Because all those who were entrusted with the management of the Nigerian state were concerned with the inheritance of the colonial policies of exploitation, they could not take advantage of the relatively 'open' economy of Nigeria to industrialize the country shortly after gaining political independence in 1960.

While it might be necessary for the state to still lead economic development in developing economies given the political economy of industrial development globally, it is important to note that, the interventionist state into the economies of Nigeria and South Korea embark more on their industrial development, but not merely on industrial growth. As noted, industrial growth can not be conflated with industrial development. In the case of South Korea for instance, while its autocratic state was able to record economic and industrial growth, in terms of development, it is yet to start. For, the kind of economic growth which the Korean authoritarian state promoted, was not a collective enterprise and therefore, led to the alienation of the majority of Koreans from the development process. All this calls for a re-visit to both the theoretical context and the policies that informed Korea's quest for industrialization. Apart from the state, another unresolved theoretical issue that has emerged from the literature, is the nature of the industrialization policies and strategies of the state in both Nigeria and South Korea.

Industrialization Policy/Strategy: To begin with, it is quite amazing that the North has so quickly forgotten its past that the only panacea it now suggests to the South for economic development, is to embrace the market. The North now insists on the market prentending to be unaware of the historical past of the South, a past that was marked by the ruthless exploitation of the latter by the former during the period of colonialism. Furthermore, its theories are rooted in the past experiences of the countries of the North with economic development. Little wonder why Africa's experiment with the development theories of the North for instance, has further deepened the underdevelopment of the continent. As former colonies, too, the Asian NICs' experiments with the development theories of the West led them into severe crises, and only began to have some growth when they abadoned them and came up with a modified version of the theories. For instance, the Asian NICs would not have experienced industrial growth if they had strictly followed the Western theories of economic development.

Yet, all this is often forgotten when most scholars compare African countries with the Asian NICs. Some of them even create the erroneous impression that, the theoretical framework that informed the path which the Asian NICs took to industrialization was built on the Western theories of economic development, whereas it was nothing of the sort. The World Bank's *East Asian Miracle* for instance, made many policy makers and scholars to believe that the rapid growth which the economies of the Asian NICs recorded in the 1980s were founded in the application of theories of the market whereas they were not. Even the studies by scholars like Amsden and Song, also made it look all the more so. The autocratic Korean state was not only in charge of the economy, but was itself, a major market force determining the direction and pattern of growth of the country's economy. All this made it difficult for the Korean economy to have self-sustaining market mechanisms that would enable it absorb major problems arising from a fast growing economy like hyper-inflation. More details on this later.

The Park regime had adopted an industrialization policy and strategy which, on paper, looked as if they were rooted in neo-liberal economic theories of the West, and in which the market is central to their successful implementation. In practice, however, they were far from being any in the sense that they underwent heavy internal modifications so that it could permit the implementation of its own export-oriented industrialization policy programmes. In other words, rather than allowing the market forces a free rein in the Korean economy as monetarist theory would dictate, the state, through its interventionist policies and projects, shaped the internal market and the pattern that industrialization later took in the country. Technically speaking, the Korean economy has, on balance, ended up being driven not by the forces of the market forces as obtained in the market economies of the West, but rather, by those of the state. Not unsurprisingly, the Korean government was pressurized by the OECD to open up its economy for the exports of its member-countries; a call the IMF has reinforced as one major conditionality for advancing a rescue loan to the Kim Dae-jung government. Partly because of the policy regime of the authoritarian Korean state, too, instead of encouraging the growth of the nascent local capitalist class which ought to have taken up the task of eventually sustaining the country's economic and industrial growth and development in the future, it nurtured the big business group with the majority of the chairmen of their board of directors as its cronies. Why did the state take such a decision? It was partly because the *chaebols* were created to serve both as the state's own institutions for lubricating its capitalism, as well as the economic vehicles for implementing its export-oriented industrialization. As a creation of the Korean state, the big trading groups benefitted from its favourable policies and incentives. For instance, the big trading groups were privileged to state-guaranteed heavy loans for which little or no interest was paid, and all this was to the detriment of local private capital. The problem was not really so much with the loan as the politics that informed the entire policy. To the trading groups, the loan was nothing less than the state's financial support to them moreso they could not really borrow from the foreign stock market. Perhaps, this explained why they were not under any pressure from the state to repay. That in part made the government-business groups relationship so inseparable that they became intricately financially dependent on the state, explaining partly why they could not withstand any stiff market competition in the local economy. As the state began its tactical divestment from the economy since it joined the OECD, most of the *chaebols* have liquidated. What is more, the Korean state had established a strong and autonomous bureaucracy which no political class could so easily sway, to implement its industrialization policies. In all this however, the state excluded labour from the political decision-making process and its industrializtion project, and this explained why the Korean labour, though cheap and relatively skilled, was still very militant and reputed for causing incessant industrial unrest in the country.

From all appearances, the path which the Korean state took to industrialization and its industrial policy, had placed the state at the heart of it with the *chaebol* as the movers. It was all prone to crisis. For instance, while the growth-obsessed Korean economy progressed, it, however, remained closed to imports of intermediate and capital goods from Western Europe including Japan and the USA. Its initial closure, as argued by the Park regime and subsequent governments in Korea, was aimed at protecting its infant industries from collapsing under the weight of dumping price of imports from the Western European countries. It was like the price that the West had to pay for using South Korea as a 'bulwark' to stem the expansion of Sino-Russian communism in East Asia. In essence, though the approach which informed Korea's industrialization did promote rapid growth in its economy, it fell far short of the principles of the market as the Bank and the Fund would want us to believe. This explains in part why caution should be exercised in comparing the South Korea's rapid economic and industrial growth with Nigeria's where the major cause of the deterioration in its economy particularly its industrial underdevelopment resides with both the nature of the state and its exploitative policies. For instance, those who were in charge of the post-colonial Nigerian state, unlike their counterparts in Korea, merely saw industrialization as a catch phrase with which to hook on to political power as all that mattered to them was how to capture and maintain state power by all means. And as a consequence, though monetarist policies largely informed Nigeria's quest for industrialization, unlike in Korea, where similar policies were also adopted in theory, but modified in practice, in Nigeria, they were left in the hands of foreign capital to execute. Would it have made any difference if the custodians of the Nigerian state have followed the path taken by the Korean path; that is, modifying the economic development theories of the West to suit the local situation?.

Not really. For, it is not as if the state managers were unaware of the implications of colluding with foreign capital to pillage the economy. Rather, they decided to do so in order to advance their parochial gains, and to the detriment of country's economy at large and the material well-being of the people in particular. But all the same, Korea's experience has shown, that the nature of industrialization arising from autocratic regimes can hardly be found in the tenets of the market. All this explains in part, why it is crucial to deeply understand the nature of the industrial policies and projects of South Korea by paying particular attention to the character of the regime under which industrialization has proceeded in order to aviod being misled into taking the mistake of taking the South Korea's impressive rapid industrial growth as a model for Nigeria. Suffice it to note that, both Nigeria and South Korea have to rethink their respective approaches to industrialization. What it all calls for, is that attempts at comparing South Korea with Nigeria in particular and the Asian NICs with Africa generally, have to be contextualized in order to avoid making sweeping generalizations or using certain basic economic indicators that might obscure the reality on the ground. The last but obviously not the least of the unsettled issues that has arisen from the received literature, is the mode of extracting surplus from the economy.

Mode of Surplus Accumulation: From the various studies that have been conducted so far on the Korean economy since its dramatic growth in the 1980s, virtually all of them evaded an analysis of its mode of accummulation. For, without a fuller understanding of Korea's production system, it will be difficult to fathom how to get the 'right fundamentals' like appropriate micro and macro-economic policy and mixes, and the conducive political environment. Nor will it be easy to create any rigorous basis for comparing Korea's efforts at industrialization with Nigeria's.

It is not in doubt that the post-colonial state in both Nigeria and South Korea is capitalist, led capitalist accummulation, and intervened into the economy most often, through its own agencies. Perhaps, that is about all that they share in common at the theoretical level, as their practical experiences differed greatly. For instance, like the Korean state, the mode of capitalist accumulation in Nigeria was state-led. However, whereas the managers of the Nigerian state merely continued with the colonial policy of industrial accumulation in which they sought accommodation with foreign capital while the country's economy was kept relatively open in order to reproduce the global capitalist system, the custodians of the Korean state closed its economy to foreign investors as it nurtured a local big business group to further its own stlye of industrial accumulation. In particular, while the Nigerian state established its public corporations, which, together with foreign capital, mediated its own pattern of accummulation in the economy including the industry, the Korean state and its *chaebols* shut out foreign capital from the internal process of industrial accummulation in the Korean economy. Furthermore, unlike in Nigeria, where the state managers were of the view that the West should, indeed, be left to define the context and content of industrial accummulation, and help implement its industrialization programmes, its counterpart in Korea adopted its own modes of industrialization and surplus extraction which were largely independent of foreign capital inspite of the fact that it was the state that attracted the funds with which it began the industrialization of the country. Inspite of the huge foreign loans that the regime of General Park attracted in the 1970s, for instance, it hardly permitted foreign capital to determine how they were spent in the Korean economy. By so doing, the state remained at the heart of industrial accummulation and ensured that, together with its local cronies, they would remain the major beneficiaries. But that also, was one of the root causes of state's incapacity to adjust the industrial sector in line with the forces of the market following the onset of hyper-inflation in the Korean economy in the late 1970s.

It is not in contention that the Korean state used the *chaebols* to promote industrial accumulation in the Korean economy. In fact, there is nothing particularly wrong with such an approach since it enabled the South Korean government to ward off foreign capital, which, if left an unfettered entry into the Korean economy, it would have brought severe economic underdevelopment to its local economy as it did in Nigeria where the state's so-called industrial policies were no less than a package of incentives to attract international finance capital into the country. However, the state/*chaebol*-led mode of industrial accumulation in Korea was ridden with contradictions which are partly manifesting in the form of economic crisis which the country is currently faced with, explaining why labour was specifically excluded from the decision-making process. This largely explains the militant nature of labour in Korea, and the deepening industrial unrest it has caused in the country since the 1970s as noted earlier.

It is not in doubt that the Korean state was all powerful. However, its mode of capitlasim still remained fragile as evidenced in its internal incapacity to accumulate without the *chaebols*, and vice-versa. All this explains in part the reluctance of the state managers to bow to the pressure of the IMF and financial institutions from the West to significantly divest from the real sector of the Korean economy especially since the advent of the recent economic crisis in 1997. Even before the onset of the recent economic crisis, the high inflationary trend that first set into the Korean economy in 1979 could not be resolved by the state largely on account of its model of surplus extraction. For instance, the state was unable to adjust its industrial policies on the heavy and chemical industrial sector in line with economic reforms that were aimed at stabilizing the economy while preventing the dominance of the big business groups. Why was the state unable to adjust its industrial sector? One of the reasons is that, the *chaebols* were so dependent on the state that they could not stand on their own if the reforms were enforced. Not only that. The managers of the state particularly the bureaucratic class, was not ready for the reforms to go through and for any change of the *status quo* since the existing mode of capitalist accumulation benefitted them. Which is all the more reason why the question is posed: Who really governs in Korea? In essence, there is a lot more problem with the nature of industrial accumulation in Korea than just the divestment of the state from the economy. It is for these and other related reasons that subsequent regimes in South Korea since the overthrow and assassination of General Park in 1979, have been unable to transcend the state/ bureaucratic/*chaebol*-mode of industrial accummulation in the country.

What all the contradictions in the Korean production system portend for Nigeria especially in terms of national industrial policy is that, though the authoritarian Korean state had promoted rapid economic and industrial growth, it did so amid inherent contradictions. Now that the basic contradictions arising from the literature have been settled, it is in order to proceed with the comparative analysis of the experiences of both Nigeria and South Korea with industrialization. How did the state approach the development of iron and steel in particular and industrialization generally, in Nigeria and South Korea respectively? This and other related issues are addressed in the next chapter.

NOTES AND REFERENCES

1. See Peter Kilby, *Industrialization in an Open Economy, Nigeria; 1945-1966*, [Cambridge, Cambridge University Press, 1969].
2. See Adebayo Olukoshi, 'The Multinational Corporations and Industrialization in Nigeria: A Case Study of Kano, C.1903-1985', Unpublished Ph.D Thesis, Leeds University, 1986.
3. Adebayo Olukoshi, 'An Assessment of the Impact of the Economic Recovery Programme of the Nigerian State on the Manufacturing Sector: A Kano Case-Study', [Ibadan: Social Science Council of Nigeria, 1991], monograph.
4. Bright Ekuerhare, *Studies in Pattern and Problems of Industrial Accumulation in Nigeria*, [Lagos: Van Hurst, 1996]. For more details, see Adebayo Olukoshi and Lennart Wohlgemuth eds., *A Road to Development: Africa in the 21st Century*, [Upsalla: Nordiska African Institute, 1995].
5. B.U.N Igwe, 'Iron and Steel and Machine Tool Industries in Nigeria', in A.Fadahunsi and B.U.N Igwe eds., *Capital Goods, Technological Change and Accumulation in Nigeria*, [Darkar: CODESRIA, 1989].
6. See Bayo Kolade, 'Capital Goods Development in Nigeria Viewed from the Perspective of the Iron and Steel Industry' in *Capital Goods and Technological Development in Nigeria*, Conference Proceedings of the Nigerian Economic Society, 1990.
7. Hatch Associate, *Steel Sub-Sector Study in Nigeria*, Report of a

Study [Toronto, 1988].
8. D.D.Dafinone & Company, *Privatization and Commercialization of Government Steel Companies*, [Lagos, 1989].
9. Omoweh Independent Committee Report, *Commercialization of the Katsina Steel Rolling Company*, [Lagos, 1992].
10. See, Nigerian Metallurgical Society's Conference Proceedings, *The Nigerian Steel Industry: Techno-Economic Appraisal*, [Lagos, 1994].
11. See, Daniel Omoweh, 'The Nigerian Steel Sector in the Global Steel Industry', *op. cit.*
12. See, World Bank, *The East Asian Miracle: Economic Growth and Public Policy*, [New York: Oxford University Press, 1993]. See also, Peter Harrold, Malathi Jayawichkrama, and Deepak Bhattasali, 'Practical Lessons for Africa from East Asia in Industrial and Trade Policies', [Washington D.C.: World Bank, Discussion Paper, 1996].
13 Sing-Young Lee, 'Country Development Experience: From Ashes to OECD', Paper delivered at the United Bank for Africa Annual International Lecture, Muson Centre, Nigeria, November 28, 1996. It was also the view of the majority of Koreans the author interviewed during his field trip in South Korea between March and August 1997.
14. See Dwight H. Perkins and Michael Roemer, 'Differing Endowments and Historical Legacies' in David L. Lindaner and Michael Roemer eds., *Asia and Africa: Legacies and Opportunities in Development*, [Carlifornia: Institute for Contemporary Studies Press, 1994].
15. Deborah Brautigam, 'What Can Africa Learn from Taiwan?: Political Economy, Industrial Policy, and Adjustment', *Journal of Modern African Studies*, 32, 1, 1994. For more details, see also, Takatoshi Ito, 'What Can Developing Countries Learn from East Asian Economic Growth?', in B. Pleskovie and J. Stiglitz eds., *Annual World Bank Conference on Development Economics*, [Washington D.C.: The World Bank, 1997].
16. See Carter J. Eckert *et al*, *Korea Old and New: A History*, [Seoul/Harvard: Itchokak/Harvard University Press, 1990]; Alice Amsden, *Asia's Next Giant: South Korea and Late Industrialization*, [New York: Oxford University Press, 1989]; and Song, Byung-Nak, *The Rise of the Korean Economy*, [Hong Kong: Oxford University Press, 1995].
17. For details, see Kyang-Hwie Mihu, *Industrial Policy for Industrialization of Korea*, [Seoul: KIET, 1985]; and Jong-Chan Rhee, *The State and Industry in South Korea: The Limits of the Authoritarian State*, [London: Routledge, 1994]. See also, Ray Kiely, 'Development Theory and Industrialization: Beyond the Impasse', *Journal of Contemporary Asia*, 24, 2, 1994.
18. Abstracted from the *Annual Reports of the Bank of Korea*, [Seoul, 1980-1992]; *Annual Reports of the Development Bank of Korea*, [Seoul, 1980-1994]; *World Development Report*, [Washington D.C.: World Bank, 1990]; and for the figures on steel, see the *Steel*

Statistical Yearbook [Brussells: IISI, 1996].
19. See Alice Amsden, *op. cit.*
20. See Byung-Nak Song, *op. cit.*
21 See Tae Wan-Son, *Development of the Korean Economy: Past, Present and Future,* [Seoul: Samhura Publishing Company, 1972].
22. Alice Amsden, *Asia's Next Giant ..., op. cit.*
23. Byung-Nak, Song, *The Rise of the Korean Economy, op. cit.*
24. Frederic Deyo ed., *The Political Economy of the New Asian Industrialism,* [Ithaca: Cornell University Press, 1987].
25. Chalmers Johnson, 'Political Institution and Economic Performance: The Government-Business Relationship in Japan, South Korea, and Tourism' in F. Deyo ed. *The Political Economy ..., ibid*; and his Japan: Who Governs? The Rise of the Developmental State, [New York: W.W. Norton & Company 1995].
26. Peter Evans, 'Coalition, Institutions, and Linkage Sequencing: Toward a Strategic Capacity Model of East Asian Development' in F. Deyo ed. *ibid.*
27. Ray Kiely, 'Development Theory and Industrialization: Beyond the Impasse', *Journal of Contemporary Asia, op. cit.*
28. Jong-Chan Rhee, *The State and Industry in South Korea ..., op. cit.*
29. Parvez Hasan, *Problems and Issues in a Rapidly Growing Economy,* [Baltimore: John Hopkine University Press, 1976], p.29.

Chapter Three

The Politics of Steel Development and Industrialization

3.1 Introduction

As its main object, this Chapter critically analyses how the contemporary level of industrialization with particular emphasis on the steel sectors of both Nigeria and South Korea came to be, taking cognizance of the role of the state. Within this context, it examines in a holistic and dialectical manner, how the forces of colonial capitalism first destroyed the primordial iron and steel cottage industries in both countries and later re-oriented them to cater for the parochial interests of foreign capital and the imperial powers. It also analyses the political environment and the regime of micro and macro-economic policies and mixes under which steel development took place in Nigeria and South Korea, especially since the period of political independence. In doing this, it examines the nature of the interplay of the policies of the post-colonial state with the forces of transnational capitalism, and in turn, how they structure steel development and industrialization in Nigeria and South Korea respectively.

To begin with, by the 1970s when both Nigeria and South Korea embarked on the development of their iron and steel sectors, the global steel industry was not already dominated by the Western European steel-producing countries like Germany, France, Belgium, Italy and Britain, and later Japan, but also, the industry was faced with the crises of overconcentration of international finance capital, closure of production lines, and declining market outlets in Europe. As the crisis in the world steel industry deepened and continued for the rest of that decade with little or no hope of recovery in sight, the European steel makers had, in an attempt to survive, taken advantage of the untapped huge steel market of countries in the developing economies which had just begun to set up iron and steel companies, to market their semi- and finished steel products, technologies and production machineries including the spares. As 'late starters' therefore, initial efforts at steel development in Nigeria and South Korea were resisted by the European steel producers. And when the latter failed to stop the former, it insisted on the relocation of parts of its 'touching-up' production lines to the developing countries. At different times in the past for instance, the West and the Bretton Woods institutions like the World Bank were vehemently opposed to the establishment of steel companies in Nigeria and South Korea, arguing that their domestic steel demand was too insignificant to warrant such huge investment. Small steel markets and low absorptive capacity were not the real reason why the Bank discouraged steel development in both Nigeria and South Korea; rather, its intention was to secure these countries as market outlets for steel exports and technologies of the steel transnationals. In South Korea for instance, the World Bank had initially campaigned against the setting up of an integrated steel plant during the regime of General Park, but when it failed, later supported the establishment of a machine tool company. So, too, did the Bank oppose Nigeria's initial attempts at developing its steel sector, prefering instead, that its local steel demand be met through imports from the Western European steel producing countries within the same period.

At its various stages of development, the steel industries of Nigeria and South Korea were also faced with difficult problems; problems which are partly contradictions arising from the global steel industry while, the rest is rooted in the approach that both countries took to industrialization. In Nigeria, for instance, its steel industry is not only still struggling to 'take-off', but also, has, from the onset, been bedevilled with deepening financial constraints, obsolete machines, lack of spare parts, and acute shortage of basic raw materials. The paradox of Nigeria's steel sector is that inspite of the abundance of basic steel raw materials like iron ore, coal and natural gas that abound in the country, the state still cannot get it really underway. For instance, before the Nigerian state embraced the adjustment package of the Bank/Fund in June 1986, the government-owned steel companies had, at various times, closed and re-opened shops due in part to their inability to source foreign exchange for importing basic steel raw materials like iron ore and billets. With adjustment, the situation worsened. Even since the official termination of the structural adjustment programme of the Bank/Fund in 1994 by the regime of the late General Sani Abacha, the steel companies were still shut with little or no hope of recovery at the time this study was conducted. Also, in the country's private steel sector where most of the companies are mini-rolling mills, production was skeletal as most of their operations were suspended indefinitely due to their inability to procure foreign exchange to import major production inputs like steel billets, a semi-finished steel product [1]. In terms of both the stage of development and nature of the crisis, Nigeria's steel sector contrasts sharply with South Korea's.

South Korea, which also embarked on steel development at about the same period with Nigeria, has been able to develop its steel sector to a level that it eventually helped catalyse the country's industrial growth. This was inspite of the initial stiff opposition from the West to its steel projects coupled with the country's lack of basic steel raw materials like iron ore and coking coal. In other words, if South Korea that lacks both iron ore and coal was able to surmount the obstacles placed on its path to steel development and recorded a relatively significant growth in its steel sector, the pertinent question that arises is: Why has Nigeria failed to get its steel industry really started?

My contention in this study is that, the real problem is not so much with the failure of the steel sector to launch Nigeria's industrial age as its development was not on the agenda of the managers of the Nigerian state. Suffice it to note that, steel development has not actually got underway in Nigeria. This explains in part why the crises in Nigeria's iron and steel sector might present themselves as economic and technological though, they are indeed, political in nature. That steel development is not really on the agenda of the managers of the Nigerian state is by no means an assumption, that efforts were not made by the government to establish public steel companies in the country. Of course, the Nigerian government has established five steel companies in the country. On the contrary, steel development as an integral aspect of the overall efforts at industrialization, has largely been one of the diversionary measures which serves as an *alibi* for the political class to hold on tenaciously to political power. Industrialisation has become a catch phrase to solicit for political support, the first being the struggle for political independence. At issue, therefore, is not that the state led steel development; rather, it is the lack of commitment and political will to embark on industrialization by those entrusted with the management of the state. For, South Korea's relatively impressive performance in steel development was also state-led. Why has the Nigerian state failed to achieve the same feat in its steel sector like South Korea? The question will be addressed in the rest part of this chapter, beginning with the role of the colonial state in the process of industrialization of both Nigeria and South Korea

3.2 Incorporation: The Colonial State and Steel Development

To start with, it was how the colonial state conceived industrialization that eventually shaped the development of the steel sectors of both Nigeria and South Korea. So, it is natural to commence the discussion with the nature of the industrial policy of the colonial state as a background context for a deeper understanding of steel development in both countries. Let us begin with Nigeria.

3.2.1 Nigeria

Nigeria's deepening industrial backwardness dates back to the period of British colonial rule. Following the 'formal' imposition of British foreign rule on Nigeria in 1900, the forces of Western imperialism particularly the colonial trading companies, destroyed the country's primordial cottage iron and steel industries notably the foundries and mini-steel works which, if they were left to grow at their own pace, would have served as a foundation for its industrialization. And as a consequence of this, the locally made hoe and other intermediate and capital goods, with which the Nigerian peasant farmer entered into colonialism were replaced with a British-made ones at the time Nigeria attained political independence in 1960. In essence, as long as the majority of Nigerian peasant farmers still depend on the imported hoe as an implement for agricultural production, their activities will remain intricately tied to the whims and caprices of the metropolitan intermediate and capital goods-producing companies, an indication of Nigeria's depeening technological dependence on foreign capital.

Why did the colonial state destroy Nigeria's primordial indutries? As a colony, the industrial policies and actions of the colonial state was not to industrialize Nigeria, but to extract and export the country's raw materials to the metropolitan countries where they were used to manufacture goods that were exported back to the country. This was one major reason why foreign capital particularly discouraged the local manufacturing of goods that were imported into the country. This was partly the origin of Nigeria's deepening technological dependence and its industrial and economic backwardness throughout the period of British colonial rule to date.[2] Since Nigeria's industrialization was not on the agenda of the colonial state, its steel sector was equally not to be developed. Let us digress a little by bringing in the experience of South Korea.

Compared with South Korea's experience which was under the Japanese colonial rule in the period, 1910-1945, there was no significant difference in the colonial state's industrial policy from that of the British in Nigeria. Together, the colonial state and the Japanese trading companies, had a common agenda in formally annexing the Korean penninsula in 1910: turning the colony into a producer of raw materials for the industry, and as a market for manufactured goods. To be expected, Korea's steel sector under Japanese colonial rule was not developed.[3] Now back to the discussion on Nigeria's steel sector.

Prior to the formal colonization of Nigeria by the British in 1900, iron smelting works and foundries among other vocations, had existed in the various parts of the territories which came to be known later as Nigeria in 1914. Notable among them was the Niger Delta. Similar pre-colonial vocations were also found among the Yorubas, who inhabited areas that later came to be known as the western region of Nigeria, and the Hausa/Fulanis who inhabited the northern part of the country. Among the people of the Niger Delta, are the Akwa, who were known to have acquired expertise in local iron smelting. For instance, the Akwa had blacksmiths who fabricated farming implements like hoes and cutlasses, and simple cooking utensils such as pots, knives and spoons using scraps of imported steel products that were first brought into the area by the Portuguese slave merchants in the 16th and 17th centuries; and more of them by the British trading companies.

As noted earlier, the goal of the industrial policy of the British was basically to extract raw materials but not to industrialise Nigeria. To be sure, the colonial state, used its policies and the agencies it created, to annihilate the primordial cottage iron and steel industries in order to guarantee the local market for the intermediate and capital goods imported from Britain and did not face any local competition. Thus, in the period, 1900-1945, the colonial state deliberately prohibited any form of manufacturing from taking place in Nigeria. This was all the more so as the British home government was concerned with the consolidation of its own steel sector as a means to rival Germany with which it was twice at war. It should be recalled that the British built its steel companies with the components of the German steel factories that it dismantled and carted away. Not only that. Though Britain had got its steel sector on stream before the world wars, it was only after the end of the World War 11 in 1945, that the government renewed efforts and rebuilt it to a level that it could be compared with other steel-producing countries in Europe like Germany which was ahead of other steel producers. All this explains in part why the British never wanted to embark on steel development in Nigeria [4].

Therefore, the view which the British took was that, Nigeria did not as yet, need a steel industry of its own. In order to achieve this policy objective, the colonial state encouraged more British trading companies to open branches in Nigeria for the distribution of British-made steel products among other intermediate and capital goods. Among the notable distributors of British-made finished steel products like steel pipes, rods and coils in Nigeria, was the United African Company Group of Companies [UAC]. Though the Group dominated the importation of intermediate and capital goods including steel products, other smaller private foreign companies of British origin such as Jos Hansen & Soehne, also imported and sold some of these goods [5].

However, by the mid-1950s when it became clear that the nationalist struggle for self-rule would, sooner than later, lead to political independence in Nigeria, the British became more concerned with how to consolidate their grip of the country's economy in the post-colonial period. For the first time since 1900, subsidiaries of the British trading companies that had been operating in Nigeria began to shift focus from purely exchange to some form of local reproduction of some of the hitherto imported light consumer goods. This was to the exclusion of the reproduction of both intermediate and capital goods and spares. This was the origin of the import substitution strategy of the colonial state, which, in practice, was merely the local reproduction of imported goods. And it was with the philosophy of import reproduction that foreign capital intervened into the country's manufacturing sector. The same strategy also informed the path taken to industrialization by the regional governments by the 1950s.[6] Let us illustrate more on this.

As it became almost certain by the mid-1950s that Nigeria would soon gain political independence, the colonial state permitted the regional governments of the East, West and North, which already enjoyed some considerable measure of self rule, to intervene into the real sectors of their local economies. With respect to steel development, they were allowed to fashion out modalities for setting up mini-steel rolling mills with annual capacities not exceeding 30,000 metric tonnes. The regional governments intervened into the industrial sector with joint venture agreement with British private companies most of which were already operating in the country. The arrangement was such that, while the foreign companies continued with the importation of steel products, it also, reproduced locally, some intermediate and capital goods. Though it looked on the surface as if the colonial state encouraged foreign capital to establish steel rolling mills, it was not so in practice as no steel company really came on stream in the colonial period. What foreign capital actually did in the non-steel sector as it shifted from exchange to the local reproduction of imported goods, was to relocate some of the stages in its downstream operations to Nigeria with all the basic production inputs imported from Europe. Not surprisingly, when it decided to intervene in the downstream sector of the steel industry in Nigeria, it reproduced basically steel coils and rods. For instance, when a faction of private foreign capital of British origin relocated from Hong-Kong to establish mini-steel companies in Nigeria in 1958, it was planned to rely on import for all their production inputs. Hence, no consideration was given to the local sourcing of the basic steel raw materials like iron ore which abound in the country. However, it was not until the early 1960s that the steel mini and rolling mills came on stream.

What all this means for the steel sector is that, virtually all steel products consumed in Nigeria throughout the colonial period, which were basically steel rods, coils and pipes, were imported from Europe; products that were used mostly for construction purposes. Even after 1960 when some mini-steel rolling mills like Continental Iron and Steel Company was established by some foreign investors of Asian origin, its product mix was still the same: rods, coils and bars. Other basic metal industries that were incorporated in the colonial period like Roadside Engineering and Foundry, established in 1952, and Crittal-Hope Nigeria Limited in 1958, also began real production in the 1960s. They, like the mini-mills, relied on imported finished steel products such as bars, flats and angles as basic raw materials for fabricating simple structural intermediate goods like steel frames for windows. From all appearnces, the colonial state was not interested in the industrialization of Nigeria including its steel sector for reasons already explained. Was the experience of South Korea different from Nigeria's?

3.2.2 SOUTH KOREA

Under Japanese colonial rule, the Korea peninsula, out of which South Korea was carved in 1945, did not experience industrialization. After almost a century of Japanese persistent economic, political and military interference in the Korean peninsula, Japan formally colonized Korea on August 22, 1910. With Japan's annexation of Korea, the Choson dynasty under which the peninsula was ruled for 518 years, came to an end. The Japanese administered Korea directly and with all seriousness. Before proceeding with the discussion, it is important to note that, though this study is focused on South Korea, our discussion on the colonial period captures the entire Korea peninsula since it was a colony of Japan. That is, the period before Korea was split into North and South after the end of World War 11 in 1945. Emphasis, however, will be more on the South. Now back to the discussion proper.

The Japanese, who were 'late arrivals' among imperialist powers, drew largely on the centuries of experiences of the Western imperial powers like Britain, Spain and France whose colonies were several thousands of kilometers away from the metropole. In order to avoid the mistake of the West, geographical contiguity, among other crucial factors, informed Japan's plan to annex its immediate neighbour, the Korean peninsula, instead of other distant countries located within the Pacific rim. The Japanese own conception of colonialism was that, the nearness of the Korea peninsula to metropolitan Japan would help reduce the financial cost of administering the colony and ease the emigration of Japanese workers and traders to the colony. Unlike the Western imperial powers such as Britain which administered Nigeria among its numerous colonies through a colonial state, Japan did not have to create an artificial colonial state to administer Korea. Rather, the imperial Japanese government established full and direct control over the ancient state of Korea with its long standing historical experience and common cultural, ethnic and linguistic ties. Why did the Japanese decide to manage Korea directly during the period of colonial period? Why was absolute power vested on the Governor-General, who also doubled as the colonial administrator of Korea?

Far from being rooted in the common heritage and shared Confucian culture on which the Japanese assimilation policy in the colonial period was built, its ultimate aim was economic. As a country, Japan lacked virtually most of the natural resources especially industrial minerals like iron ore, limestone and ferroalloys that could facilitate its industrialization. Additionally, the pre-1945 Japanese economy was based largely on exchange. All this made the Japanese trading companies to mount pressure on the Japanese government to pursue an expansionist policy within the sub-region. In fact, the annexation of the peninsula by the Japanese government was partly an attempt to meet the trading companies' increasing quest for larger market outlets for their manufactured goods and guaranteed sources of raw material outside of the mainland Japan. All this accounted largely for Japanese several military incursions into the Korea peninsula before it was finally colonized in 1910. It was also partly in its search for raw material sources that led Japan to fight numerous wars with China and Russia especially to gain control of Manchuria which was rich in both industrial and strategic minerals. In essence, notwithstanding the different methods of administering their colonies, the ultimate aim of the colonial policies of the British and the Japanese in Nigeria and Korea respectively, was basically economic.[7]

After the fall of the Choson dynasty in 1910, Japanese private and public capitals took full advantage of pre-colonial agrarian Korean economy and exploited its vast natural resources like gold, silver, iron ore, tungsten, and timber. Most of these natural resources were found in the northern part of the peninsula which, by 1945, became the present-day North Korea. Chinese companies, too, were not left out in the exploitation of Korea's natural resources. As noted earlier, this accounted in part for most of the Sino-Japanese wars before 1910. Since 1910 when the Governor-General took full control of Korea, the colonial policy conferred the exclusive exploitation of its natural resources on Japanese companies. This was similar to the policy of British colonialism in Nigeria. Unlike in the north, the southern part of the peninsula was barely endowed with industrial minerals and natural resources. Rather, it had a flourishing rice milling industry which Japanese companies also later dominated thereby reducing Koreans to labourers. In essence, the northern part of the peninsula had a concentration of mining companies, which, in turn, laid the foundation for its relatively more developed industrial base than the southern part at the end of Japanese colonial rule in 1945. What the southern part of the peninsula was noted for, was the cultivation of rice and other crops like tungsten. Like the British, these natural resources were exploited by Japanese companies for onward shipment to the factories in metroplitan Japan where they were processed and re-exported.

Unlike in Nigeria where the intensification of the nationalist struggles in the post-1945 period showed that, sooner than later, the country would become politically independent, forcing the British companies to shift slightly from purely exchange to the local reproduction of hitherto light consumer goods, the

Japanese were undaunted by the agitations of Koreans for self rule. In fact, never a time had it crossed the mind of the Japanese that Korea would ever become politically independent. This was to influence the structuring of the colonial economy of Korea by the Japanese. By the 1930s for instance, Japanese companies had relocated some of their 'touch-up' operations to Korea ostensibly for the local reproduction of light consumer and intermediate goods in the peninsula. Like the British, the ultimate aim of the Japanese in doing this, was not to industrialize the colony. What took place rather, was merely a transfer of certain stages in the manufacturing industries and the semi-processing of raw materials to Korea in order to complement the operations of the manufacturing companies in mainland Japan. In other words, while the end-product activities were relocated to Korea, the real manufacturing companies where the technology also resided, still remained in the metropole. This partly accounted for why over 70 percent of both agricultural products and semi-finished products of Japanese heavy chemical industries that operated in Korea in the 1930s exported their products to the homeland Japan, the remainder was utilized by other Japanese companies.[8]

One of the major consequences of Japan's colonial policy was the exclusion of Koreans from participating in the real sector of the economy. In commerce where Koreans were allowed limited entry, they had weak capital base. This eventually put the Korean local business class at a disadvantage in the sense that it could not engage in real production much more competing with the Japanese companies. Rather, all that the Korean merchants could effectively do, was to engage in the distribution of Japanese imported goods. As the distributive trade progressed, some Korean merchants were, however, able to build a capital base that could enable them to venture into light production, particularly in the textile industry for which vast past experience already existed. Notable among the Korean companies that engaged in textile production, were Chongno Textile Company and Kim Tok-Chang. Both companies had tried without success, to break the near monopoly position which the Japanese trading companies like Mistui and Mistubishi held in the textile industry of Korea during the colonial period. They failed not strictly because of lack of expertise; rather, as part of the Japanese colonial policy of 'non-industrialization', they were starved of credits by the Japanese banks. Not only that. As part of Japan's general colonial policy, certain Japanese banks were specifically established to extend loans only to Japanese companies operating in Korea. Notable among them was the Daiichi Ginko, which later metamorphosed into the Bank of Choson, and in that capacity, played the role of the central bank for the colony. Shokusan Ginko, a Japanese industrial development bank, dominated the finance industry of colonial Korea. With the firm control of the finance sector and the Korean economy in general under Japanese banks, whatever major industrial investment that took place in the peninsula, was aimed at integrating its economy deeper into the capitalist system of metropolitan Japan.[9] Compared with Nigeria's experience under British colonial rule, there was really no much difference both in terms of the domination of the economy including its financial sector and the policy regime that governed the granting of credits.[10]

By the late 1930s, the Japanese were almost certain that Korea was better left as an outpost for raw material and market since its industrialization was not on their agenda. As a result, rich as the northern part of the peninsula was in iron ore, there were no steel companies established to consume it. Throughout the colonial period, the Japanese were not interested in developing Korea's steel industry. Rather, what was of paramount interest to the Japan Steel Corporation whose subsidiary exploited iron ore in Korea, was to semi-process the ore before shipping it to its steel companies in metropolitan Japan where it was used to manufacture steel products. The same was true for the production of pig iron and semi steel products like billets in Korea during the period of Japanese colonial rule. Of the 59,000 tonnes of pig iron produced in Korea by Japanese companies in 1934 for instance, all was exported to Japan for consumption by its steel companies. And out of the 100,000 tonnes of steel products that were cast by Japanese-owned foundries in Korea within the same period, all was exported to the homeland.[11]

Rather than deteriorate with the outbreak of the World War 11 in 1935, a war in which Japan fought, but lost, the Korean economy still witnessed some measure of growth which arose largely from the activities of the Japanese companies, and not that the economy was really self-propelled. As noted earlier, there was some slight structural shift in the Korean economy from purely exchange to the local reproduction of some of the Japanese imported goods as the British did in Nigeria. For instance, the share of manufacturing in the industry increased from 17.7 percent in 1931 to 40 percent in 1939. Within the manufacturing sub-sector, chemical, tool and other metal-based industries witnessed some increases.[12] However, the so-called growth which the Korean economy witnessed in the later part of Japanese colonial rule was, indeed, partly its response to the economic growth of Japan. For one thing, not only was the Korean economy under the firm control of Japan, but also, the agents of the growth were no less than Japanese companies. All this furthered the incorporation of the Korean economy into the Japanese capitalist system. This structural imbalance was quite evident in the pattern of Korea-Japan trade: 95 percent of all Korean export went to Japan, while 80 percent of imports to Korea came from Japan.[13]

As the World War II raged, the Japanese still continued with their industrial accumulation in Korea not necessarily because they were on the winning side. On the contrary, Japan kept most of the industries in production in order to meet its defence needs particularly for the war purposes. Therefore, the expansion which the chemical industry witnessed by 1940 for instance, was not so much an index of an industrial growth, but rather, to provide munitions for Japan's army. The same was true for the machine and tool companies which manufactured war equipment and their parts; and the textile industry that provided uniforms and clothings for the Japanese armed forces. Since most of these industries were located in the northern part of the peninsula, that is, the North Korea of today, they helped lay the foundation for its present relatively more developed defence industry compared with South Korea in the post-colonial period. As an agrarian economy under Japanese colonial rule, South Korea was industrially underdeveloped by August 1945 when it became politically independent.

As a colony of Japan, the relatively increased structural transformation that Korea's economy witnessed particularly during the period of the World War II was largely tailored to meet Japan's war needs. And this should not be construed to mean that the Japanese relocated some of their manufacturing companies in an attempt to industrialize the peninsula; rather, it was to enable Japan to prosecute the war. As noted earlier, Japan had annexed Korea in search of secured raw material sources and market outlets for its manufactured goods. Therefore, with the signs of an imminent defeat of Japan in the war becoming more real by early 1945, Japanese private capitals, became more apprehensive of the security of their investments in the post-1945 Korea. Since Korea was planned to be fully integrated into metropolitan Japan, the Japanese imperial government did not create a local business and political class to protect its economic interest, nor could it create one by 1945 and suffering a defeat. Nor to create an *ad hoc* local political class that would help protect their investment in the post-colonial period of Korea. This was in sharp contrast with Nigeria where the British used 'divide and rule' tactics to govern the colony while it created a pliant local political class that protected its economic interests while contuining with its exploitative policies and structures in the period after 1960. In essence, with the defeat of Japan in the World War 11 in 1945, and the termination of its colonial rule in Korea in the same year, most Japanese investments were seized by the Korean governments partly as a retaliation for the long period of ruthless exploitation under Japanese colonialism. That is not all. Rather than pull out of the part of the peninsula that was later known as South Korea after the end of hostilities in 1945, the armed forces of the United States of America that fought the Japanese during the war, even stayed back ostensibly to checkmate the communist incursions of China and Russia into the country; and as well, to forestall Japan's future expanionist policy in the peninsula. It was these and other related social, economic and political forces that have come to influence the creation of the post-colonial state in South Korea as well as its industrial and economic policies.

One of the main issues that is quite obvious from the colonial industrial policies of both Britain and Japan was that, steel development was not on their agenda in Nigeria and South Korea respectively. Therefore, the development of iron and steel industry in both countries only began in the post-colonial period. It was even embarked upon by the post-colonial state moreso as the local business class had a weak capital base and could only intervene into petty trading. Since the state pioneered the development of the steel sector and industrialization generally in both countries in the post-colonial period, it is in order to examine the approach they took to iron and steel development, taking cognizance of the nature of the state and its macro and micro-economic policies and mixes, and the global context of the steel industry in which it all operated.

3.3 THE POST-COLONIAL PERIOD: THE STATE AND INDUSTRIALIZATION

Why was the development of iron and steel industry state-led in both Nigeria and South Korea? As noted earlier, it was partly because the forces of colonial capitalism never really had plans to industrialize the colonies beyond using them purely as sources of raw materials and as market outlets for manufactured goods. In an attempt to reverse the inherited industrial backwardness of both Nigeria and South Korea, the post-colonial states intervened into the development of iron and steel among other strategic sectors of their respective economies. There were differences in the approach they took to steel development, though it was all state-led. Let us begin the discussion with Nigeria.

3.3.1 Nigeria

In Nigeria, the managers of the post-colonial state are not particularly different from their predecessors in the sense that they, too, adopted the import substitution strategy on which the colonial industrial policy of the British was hinged. Adopting the import substitution policy only helped deepen the integration of Nigeria's economy into the British capitalist system as hitherto imported goods were only reproduced locally. Even then, it did not bother the custodians of the Nigerian state as long as it benefitted them.

With respect to steel development, those in charge of the state had embarked on steel development pretending to be following the path which some of the major steel producing countries in Western Europe like Britain, Germany and France had taken while developing their steel sectors. In reality however, their intention was to use it to amass wealth illegally. This defensive radical posture of the state was quite evident in its position on steel development. For instance, the Nigerian state is of the belief that, the steel sector is too strategic to the economic development of a country to be left in the hands of private capital, local or foreign; and that, the huge capital outlay required for steel development was beyond the capacity of local private capital, which at the time Nigeria gained political independence in 1960, was even too weak to embark on such project. Yet, it adopted the import substitution strategy and left its implementation with foreign capital. It did so because of the rent that would accrue to its members from the importation of production machineries and their spares. Hardly any thought was given to the development of the local capacity to endogenize the technologies embedded in these goods by way of setting up steel companies with facilities for producing flat steel, and foundries where the major part of machines could be fabricated, and put in place facilities that can duplicate them and their parts. In essence, the state approached steel development in Nigeria with the hope of relying solely on the importation of the production machineries, their spares, raw and semi-finished steel materials. Little wonder the steel companies were assembled on turn-key basis by Western European steel transnationals. A brief history of steel development in Nigeria will shed light on this.

During the first decade of Nigeria's political independence, the government did not consider it particularly necessary to intervene in the development of iron and steel in the country. Why? It was the view of government that, with a *per capita* steel consumption as low as 25kg, it was better to encourage both local and foreign private capitals to establish mini-steel rolling companies to complement the importation of finished steel products. As a result, the few mini-steel mills that came on stream, had installed annual capacities not exceeding 100,000 tonnes and they reproduced basically structural steel products like bars, rods, wires and coils to supplement imports. The emphasis on the local reproduction of simple structural steel products arose from the need to construct, in the post-independence period, more social infrastructures like roads, public buildings and hospitals for which steel products were used.

The first private steel company in Nigeria was the Niger Steel Company, Emene, Enugu, which was established in 1962. The company was jointly owned by then Eastern Regional Government [later East Central State] and Ferrostaal, a German group of private capitals. The Niger Steel Company was a joint venture in which the Eastern Regional Government held 50 percent and the remainder was controlled by Ferrostaal. As the technical operator of the joint venture, Ferrostaal provided the technology and supplied the steel billets, which was the company's basic production input. The company however, suffered serious setbacks the most important of which was the collapse of its mill during trial production. Official sources said that, the basic cause of the break-down of the rolling mill during test-run was traced to the use of refurbished parts from a cannibalised steel mills in South Africa. Indeed, the trend since the 1960 is for the foreign steel companies to scrap aged steel mills and use their parts to build steel companies in Africa instead of manufacturing new production machineries. The collapsed rolling mill of the company was yet to be repaired before Nigeria was engulfed by a 30-month civil war in 1967. Even by January 1970 when the war ended, the state government and the German technical partner did not make any effort to rehabilitate the company. In 1972, it finally folded up.[14] However, more private steel companies came on stream.

In 1968, some foreign investors of British origin relocated from Hong Kong to establish the Continental Iron and Steel Company [CISCO] at Ikeja-Lagos. The company was initially wholly owned by the British. By 1972, however, its ownership was split in a ratio of 52:48 between the British and Nigerian private investors as a result of the Nigerian Enterprises Promotion Decree of 1972 [Indigenization Decree] as amended in 1977 [15], and was repealed in 1997 by then Abacha administration. Like Ferrostaal of Germany, the foreign partner, which was of British origin, did not only build and maintain the rolling mill, but also, supplied the steel billets which were all imported from Europe.

In 1970, that is, two years after CISCO was incorporated, another group of British private investors also relocated from Hong Kong to establish the Universal Steel Company [USC] at Ikeja-Lagos. Like CISCO, ownership of USC was first wholly British before it was shared between the British and Nigerian private investors by 1972 in line with the Indigenization Decree. The company was also built to depend both on foreign technology and imported semi-finished steel raw materials. Other private steel companies that were established later, were not different in the sense that they all depended on foreign steel companies for both technology and raw materials. Table 3.1 presents a clearer picture of the major private steel companies in Nigeria, their product mix and status at the time this study was conducted. Locate table 3.1 at the end - Statistical Appendix.

One major issue that can gleaned from Table 3.1, is the dominance of the private steel sector by foreign capital most of which were of British origin. Why was it so? It is partly because of the legacy of the British colonial rule. As part of the British colonial policy, it was only companies of British origin that were allowed to do business in Nigeria, then a colony of Britain. Although, after the attainment of political independence in 1960, the Nigerian government had flung the door open to investors from other European countries other than Britain, preference was still given to British foreign capital due in part to the inheritance of a political class and bureaucracy that were created to protect British economic interests. All this explains the continuation with the colonial exploitative policies and structures in the post-colonial period. Suffice it to note that, steel capitals from other European countries like France, Austria, Belgium and Russia; and others from the Asian/Pacific region like Japan, were relatively 'new arrivals' most of which came in the 1980s when the Nigerian government built five steel companies.

As noted earlier, the overall level of steel consumption in Nigeria in the period, 1960-1965, was quite low that it did not warrant the intervention of the state. After the end of hostilities in 1970, there was a sudden increase in the demand for steel products locally. The reasons were not far to seek. First, there was need to undertake the rehabilitation of infrastructure in the war affected areas in the country most of which were in the eastern region. Second, with the creation of 12 states in 1976 out of the old four regions, and later increased to 19, 30 and now 36 states, more social infrastructure like roads, public buildings and stadia had to be constructed and for which a lot of steel products were needed. Third, there was a significant increase in the development of private property. All this raised the local steel consumption per head from 30kg to 150kg. The problem was not so much with the dramatic rise in the local steel demand as the capacity of the domestic steel companies to meet the increase. For, out of the overall 150kg of steel consumed per head in the 1970s, the local production capacity could only satisfy a mere 20kg, and half of this even came in as imports.

As the demand for steel products rose dramatically in the 1970s, the thinking of the state then, was how to increase the local capacity to produce steel products for the country's domestic need. Faced with the limited capacity of the mini-steel mills to cope with the increasing local demand for steel products, the government finally intervened in the steel sector by establishing its own iron and steel companies. Unlike the private steel companies that were both technologically and raw material dependent on foreign capital, those established by the national government were in theory, planned to rely on iron ore, coal, limestone among other basic raw materials that abound in the country. In practice, it was not done since the steel companies depended on imported raw materials. As previous efforts by the eastern regional government to establish its own steel companies were beset by lack of finance and stiff opposition from the Western countries until it reached an agreement with Ferrostaal; so, too, were initial attempts at steel development by the government in past, met with lack of fund and resistance by the European steel producing countries like Britain and the Bretton Woods institutions such as the World Bank. For instance, the World Bank had, in the past, frustrated initial moves from the Nigerian government for financial assistance to etablish integrated steel companies. What the Bank and the European steel producers have always wanted for Nigeria, was the establishment of mini-steel mills and a machine tool company. Why? Since they were planned to be import-dependent for both raw material and technology, Nigeria would remain as a major market outlet for steel products and technologies of the Western steel transnationals.

However, it was only in the 1970s that the Nigerian government's previous efforts to set up its own iron and steel companies were translated into action due largely to the windfall from the oil boom that started in 1973. For instance, though the government of General Yakubu Gowon had created the National Steel Development Authority [NSDA] in 1970 to carry out a feasibility study on the possibity of setting up an integrated steel company in the country, it was not until the late 1970s that the Authority began to play some role in the development of the country's iron and steel industry. For instance, the NSDA invited British, French and American steel firms to carry out geophysical mapping of the country as well as delineating its iron ore - bearing areas with a view to deciding where to locate the proposed steel companies. Together, they all turned down the Authority's invitation not necessarily because Nigeria had, at the time, lacked basic infrastructural amenities like good network of roads, stable power; and faced with problem of a poor industrial base as they would want us to believe. On the contrary, though these problems were there, the real reason was the opposition by the steel producing countries of the West and the Bank/Fund, to preserve Nigeria more as an outlet for marketing their finished steel products, obselete production machineries and source of steel raw materials, but not to be a steel producer.

With the rejection of its proposals by the West, the NSDA was compelled to invite the Techno-Export [TPE] of the Soviet Union to carry out both geological and geophysical mappings of the country and identify the availability of the basic raw materials for the steel industry in order for it to take off. All this compelled the Nigerian government to turn to the then USSR, which as a socialist country, was not a 'traditional' friend of Nigeria with all the ethos of capitalism. So, in 1970, the TPE signed contract 1717 with the NSDA. The entry of the TPE into the Nigerian steel industry was not received by the West whose main fear was that, the steel project would provide an opportunity for Moscow to introduce and consolidate its communist ideology in Nigeria, a country with about the largest steel market in Africa. Furthermore, the signing of contract 1717 triggered off stiff pressure which the Western European steel transnationals mounted on the Nigerian government. Individually, the steel transnational struggled for contracts to build at least one of the five public steel companies. The Japanese foreign investors who were relatively new in the Nigerian economy compared with those from the West, were not left out in the scramble for contracts in the steel sector. Each of the foreign steel companies brought before the Nigerian government, the recent steel technology it had and the advantages it offers such as time of construction and maintainance. All this helped internationalize the politics of steel development in

Nigeria, and how it was later played out in the country, particularly the path that the state finally took to the establishment of the public steel companies, the technological route, and who got what contract and at what cost.

As the government's efforts at steel development progressed in the late 1970s and early 1980s, more institutions were created out of the NSDA especially to cater for and complement other aspects of the iron and steel industry. For instance, the exploration and mining unit of the NSDA which was moved to Itakpe, and later renamed the National Iron Ore Mining Company, was responsible for mining iron ore for use at the Ajaokuta Steel Company. Also, NSDA's mineral processing unit was relocated to Jos, and later became known as the National Metallurgical Development Centre, Jos. Among its duties, was to characterise the grades of various iron ore deposits and other steel raw materials like coal, limestone that abound locally essentially for the purpose of using them at both DSC and Ajaokuta. By 1979, the NSDA itself, was dissolved and replaced with the National Steel Council, which was later renamed Steel Raw Material Exploration Agency [SRMEA]. The SRMEA was charged with the responsibilty of exploring basic steel raw material, that is, iron ore, for Nigeria's steel sector. Last but not the least of them, was the Geological Survey Deparment whose major duty was to provide geological and geophysical mapping of the country's iron ore deposits among other steel raw materials.

The thinking of government in splitting the NSDA into four specialised units was that, while encouraging specialization in the various sections of the steel industry, it would expedite the development of the country's steel industry. Unfortunately, it turned out to be counter productive in the sense that, rather than complement, it engendered undue rivalry between and among them. It was a rivalry that stemmed in part from the government officials' desire to use their official positions to amass wealth particularly as that has been the norm rather than the exception in Nigeria. For instance, instead of the SRMEA sending samples of steel raw materials to the NMDC at Jos for processing and characterization, it decided to undertake the task alone. To avoid being seen as redundant, the NMDC, in turn, was compelled to extract samples of minerals for processing. They hardly exchanged ideas on their findings. Worst still, these government agencies have developed into full government parastatals, but all accountable to the supervisory Ministry of Power and Steel. These were some of the basic contradictions inherent in the state's approach to development of the country's steel sector.

As explained earlier, the government created the NSDA to quicken the construction of the Ajaokuta Steel Company. It was no surprise therefore, that the National Steel Council, predecessor of the NSDA, surveyed the initial site where the Ajaokuta Steel Company was established in 1978. When the state intervened in steel development, its plan was to establish one integrated steel company, and that was the Ajaokuta Steel Company; hence it was installed with a capacity that would meet the country's steel need. In essence, on paper, all steel-related government agencies that were created since the NSDA came into being, were expected to work towards the successful completion of Ajaokuta. In pracrice, it hardly worked as planned for reasons already explained.

Though, the Murtala/Obasanjo administration had taken more concrete steps towards steel development by signing all the contracts for the public steel companies before its exit in 1979, civil works did not begin at Ajaokuta until 1980 few months after the Shagari-led civilian came into power. Being military and poised to hand over the reigns of power to the civilian, the thinking of the Murtala/Obasanjo regime was that, the greater and real part of the development of the country's steel sector was better left for the in-coming civilian government to handle.

Even by 1979, the global iron and steel industry was already immersed in severe recession due largely to the overcentralization of steel capitals and technologies, shut down of production lines, down-sizing, declining returns on investment, and poor price regimes. In effect, the Shagari-led regime inherited a steel project that was embarked upon by its predecessor at a time when the world steel industry was already in structural crisis. As the recession in the global steel industry deepened, the European steel transnationals competed grimly between and among themselves as they lobbied the Shagari administration for contracts for the construction of government-owned steel companies. With the TPE assured by the government of the contract for the construction of the blast furnance at Ajaokuta Steel Company, Euro-American steel capitals had, together with

Japanese steel companies, mounted spirited campaign and lobby to get a share of the sumptuous steel contracts in Nigeria. First, the West had told the officials of the government that, the Soviet steel technology was obsolete, recommending the automated and turn-key steel technology which they have, as modern. Second, the Western European steel transnationals also tried to convince the government that, the direct reduction steel technology which they have, would take a maximum period of 30 months to build as against the blast furnace technological route of the Soviet that would be completed in at least six years. Severe as the external pressures on the government were, it did not really have steel development on its agenda as explained. Perhaps, this partly accounted for its inconstitency in the path it took to steel development. In other words, the government's preference for simple structural steel products like bars, coils, and rods was merely a cioncidence with the interest of the Western steel companies. As it shall be made clear in the next chapter, the monoproduct mix of the steel companies only helped perpetuate the country's technological and industrial underdevelopment at the time when South Korea was able to use its steel sector to partly reverse its industrial backwardness.

The fact that those in charge of the state were not really interested in steel development not only trivialised issues such as where the steel companies were to be located, but polarised them along the North-South dichotomy that has bedevilled Nigeria since tha colonial period. For instance, the location of Ajaokuta Steel Company at Ajaokuta, an Igbira community in the present-day Kogi State, had generated sentiments among the political class from the South, who felt that it was a continuation of the overall design by the North to continue to marginalise the southern part of the country. In essence, the Ajaokuta Steel Company was understood by the people from the southern part of the country as exclusively meant for those from the north. Even those from the northern region of the country saw it that way. All this reinforced the pressure of the southerners on the government to establish another integrated steel plant for them. The politics of citing the steel companies was taken full advantage of by the Western steel transnationals to intensify their campaign for the establishment of additional integrated steel company in the country even when it was not economically wise for the government to embark on such ambitious steel project. In any case, the state managers and their allies saw the steel projects as yet another avenue to amass wealth

Beside the politics that surrounded steel development, the state's own mode of accumulation in the economy significantly structured Nigeria's contemporary steel sector. As noted, those entrusted with the management of the state has been more concerned with the consumption of public wealth accruing largely from oil rent and not its generation. Thus, the development of steel provided yet another avenue for them [custodians of the Nigerian state] to quickly grab their own share from the national wealth most of which came from oil production and export. Not surprisingly, either they used their political positions to corner contracts, or acted as fronts for the steel transnationals. In fact, amassing of wealth using political power instead through enterpreneurship, was only aggravated when the state embarked on steel development. Unsurprisingly, Mallam Ali Makele of blessed memory, who was then Minister of Steel under the Shagari regime ensured that 'steel flowed' since the steel project became the 'catch phrase' of the regime to justify its ambitious budget in the face of dwindling funds accruing to the state from oil export.[15] Ensuring that 'steel flowed', did not in any way mean that the state was really committted to Nigeria's industrialisation. On the contrary, what it represented was really no less than a defensive radical posture by the state managers pretending to be concerned with the industrial development of the country whereas their real intention was to use the steel project, like the establishment of petroleum refineries, as an avenue to accumulate moreso since it was a relatively 'new project'. As shall soon be revealed in the course of the discussion, the unseriousness of the government with steel development in particular and industrialization generally, is quite evident in the overall structure and performance of the steel companies and their linkages with private steel sector. Not only that, as there were glaring evidences of overinvestment on the public steel companies in comparison with the cost of establishing similar steel companies and of the same capacities elsewhere in the developing economies.

Between 1979 and 1983, the government started and commissioned the Delta Steel Company [DSC], Aladja-Warri; three inland steel rolling companies at Katsina, Jos, and Oshogbo; got civil works underway at Ajaokuta, and began to instal facilites that would enable the National Iron Ore Mining Steel Company located at Itakpe, to commence production. However, since the steel project was only important to the state managers in so far as it permitted them to consolidate their accummulative base, they showed little or no concern for the structure of the steel companies and in particular, their product mix. For one thing, the capacity of the steel sector of a country not the least Nigeria, to act as a bedrock for its industrialization is intricately tied to the nature of the products rolled out by the steel companies. Yet, the government biased the product mix of the public steel companies for simple construction such as steel rods and coils; the same products that were already rolled out of the mills of the private steel companies. According to government, given the poor industrial base of the country at the turn of the 1970s, it was considered a more important priority to first meet the local steel demand both for the construction of social infrastructures and property development before considering the local production of capital goods. Again, it was all a farce. This was compounded by the nature of the political economy of the global steel industry which foreign steel capital had dominated as well as the steel market of the developing economies like Nigeria.

As a consequence of all this, Nigeria's local steel need was still largely met through imports even long after the public steel companies had begun operation. This was all the more so because of the state managers' penchant for contract especially as it held promises for immediate self-enrichment. Hence, it did not really matter to the state why Delta Steel Company was commissioned with imported iron ore inspite of its abundance locally. That is not all, as barely six months after its commissioning in 1982, the company was bedevilled with financial crisis so much so that it could not procure its production inputs. The three inland steel rolling mills located at Katsina, Jos and Oshogbo were planned to depend 100 percent on DSC for their billet needs. Because DSC could not even provide 10 percent of their billet need, the rolling mills were marred by undercapacity utilization right from inception. Table 3.2 presents a clearer picture of the nature of the public steel companies. Details on the performance of the steel companies are treated in the next chapter. Locate Table 3.2 in the Statistical Index.

At issue, is not whether the state has established steel companies; rather, it is the extent to which they have been used to reverse Nigeria's industrial underdevelopment. Which is why it will be misleading to blame the North for Nigeria's unimpressive attempt at steel development and its deepening industrial backwardness. Let us now discuss South Korea's approach to steel development.

3.3.2 South Korea

To begin with, South Korea, unlike Nigeria, was not endowed with basic steel raw materials like iron ore and coking coal. That explained in part the difference in the approach the Korean state took to steel development. To an extent, however, both countries still have one thing in common and that is, steel development was led by the state in the post-colonial period. But that is about all that they share in common since the managers of the Korean state were markedly different in both political will and commitment to steel development from their counterparts in Nigeria for reasons already explained. Our concern here, is not to repeat these reasons, but to bring into bolder relief, how the differences in the approach the state in Nigeria and South Korea took to industrialisation have come to explain the sharp constrast in the level of development of the steel industries in the two countries.

As noted earlier, the Japanese had no plan to industrialize South Korea during the colonial period. So, it would hardly have made any difference if Korea were endowed with iron ore and coal.

Though the United States of America government's real interest in the peninsula was to use South Korea to contain the spread of Sino-Russian communism in the sub-region, it was the protection of this strategic/political interest that also helped prevent Korea's nascent capitalist class from being wiped away by the politics of liberation. Unfortunately, however, the nascent local capitalist class in South Korea was not oriented towards production, but on exchange. In essence, Korea's local capitalist class, like its Nigerian counterpart, was so weak that it could not be relied upon to take up the development of the country's steel sector and its industrialization. As noted, given circumstances that led to the fromation of the government of Syngman Rhee, it could hardly do much to reverse Korea's severe economic crisis and industrial backwardness which were inherited from Japanese colonialism. For instance, until 1951 when North Korea went to war with its neighbour in the South, the Rhee's government was unable to lay serious foundation for the industrial development of the country. And even after the end of the war in 1953 and until 1961 when his regime was overthrown in a *coup d'etat*, Korea's post-colonial economy was still far from being put on course as poverty and hunger deepened. For a country that was so poor and battling with internal basic social problems like hunger, illiteracy, lack of pipe-borne water, it was difficult to expect the Rhee's government to think of asking for aids with which to embark on the country's industrialization including its steel sector. In any case Korea's industrialization was not on the agenda of the Rhee's administration. The Western European aid donor-countries were themselves, not really interested in advancing loans and technical assistance to South Korea for tackling the fundamental problems that the country was faced with; that is, economic and industrial underdevelopment. Let us dwell more on this with emphasis on the steel sector.

By 1945, South Korea had really no steel industry of note due in part to the industrial policy of the Japanese colonial government. What sprang up after the end of the North-South Korean war in 1953 instead, were some mini-steel mills that were privately owned. Notable among them were: Inchon Iron and Steel Company, and Union Steel Manufacturing Company, both were engaged in the production of simple structural steel like rods, and forging of basic intermediate goods. They had a combined annual capacity of 150,000 tonnes. One of the reasons for their small capacities stemmed from the insignificant level of a domestic *per capita* steel consumption, which was estimated at 20kg. They were also 100 percent dependent on imported billets with their product profile basically made of steel rods and coils.[16] From all appearances, Korea's steel sector was as underdeveloped as Nigeria's within the first two decades of its political independence.

As explained earlier, the majority of Koreans felt that, the fundamental problem that South Korea was faced with, was economic and not really political, whereas the root cause of the problem was political. That partly explained why after the overthrow of the Rhee's government through a *coup d'etat* led by General Park Chung-Hee in 1961, the Park regime embarked on the project of re-building Korea's political economy but with a high premium placed on the economy. In order to overcome Korea's economic backwardness, the need arose for the post-colonial Korean state to be reconstituted. That accounted for the paradigm shift in the reorientation of those who were later entrusted with the management of the state and the Korean economy. As part of the general consciousness for change, the military government of then General Park became concerned with how to reverse the country's deepening economic underdevelopment as well as developing the capacity to contain the security threat posed by its neighbour, North Korea. The regime was of the belief that it was mandatory for all those entrusted with the management of the Korean state to undergo a rethink of the path that its predecessor took to economic development. An important part of this new thinking, was Korea's industrialization, which to Korean policy makers, was also seen as perhaps the only viable means through which the country's chronic poverty and the overall socio-economic backwardness could be reversed. As part of the government's drive for industrialization, it placed high emphasis on steel development. In fact, the Park regime saw steel development as a 'national duty' in which all Koreans especially those at home, were expected to be fully involved. The major industrial policy of the Korean state was basically to transform South Korea into an industrialized country with strong emphasis placed on export promotion.

As part of Korea's general industrialization policy and programmes, the heavy and chemical industries of which the steel sub-sector was an intricate part, were exclusively left for the state to develop. Why? Those in charge of the Korean state were of the belief that, as a 'late industrializing' country lacking basic steel raw materials, the steel sector was not only regarded as the pillar of the country's industrialization, but that its development at the early stages, was better left for the state to handle. All this made steel development central to the industrialization policy and project of the then government of General Park, which subsequent regimes have followed with little or no amendment as of 1998. The steel policy of the Korean state was based on sustainable and balanced industrial development and technological autonomy. Within this framework, it was envisaged that the country's industrial and technological growth would ultimately be structured by the products of Korea's steel companies.[17] This contrasted sharply with Nigeria where there was yet to be a steel policy at the time this study was conducted. It is not that the government which is largely in charge of steel development in Nigeria is not aware of the advantages of having a steel policy in place. Rather, the state, is itself, not really interested as noted earlier. Unlike in Nigeria, too, which is well endowed with steel raw materials coupled with the huge revenue accruing to the state from oil production and export, yet it still failed to get its steel sector really underway, South Korea's relative poverty in all this, did not impede its effort at steel development. The pertinent question is: How then, did the Korean state embark on steel development?

The development of steel in South Korea did not start until 1968 due not so much to the lack of steel raw materials, but largely to the lack of finance. As a relatively poor country in the 1960s, the then regime of General Park only embarked on the country's industrialization hoping to get funds from external sources. With the opposition from the West and the Bretton Woods multilateral financial institutions like the World Bank and the International Monetary Fund to Korea's request for financial assistance to establish its own steel companies, it was almost certain that Korea's steel project would hardly get underway. In the face of such opposition, it became clear to the majority of policy makers in Korea and scholars as well, that the West was only interested in using the Republic to contain Sino-Russian communist expansion in the region. Japan took advantage of the rebuff of the West to reconcile with South Korea, its former colony. The reconciliation took largely the form of the payment of reparation to the Koreans who were ruthlessly exploited by the Japanese in the past. It was with the $300million paid by the Japanese government as reparation in 1968, that South Korea under the then administration of General Park, began the development of its steel sector.

As a developing economy, the regime of General Park had, shortly after assuming office in 1961, adopted the import substitution strategy the following year as Korea's industrialization policy and strategy. This is inspite of the failure in the majority of the African and Latin American countries where the import substitution strategy was experimented with. South Korea's approach to import substitution was different from that in Africa in the sense that its implementation was controlled by the Korean state as opposed to leaving it for foreign capital to execute, the situation in African and Latin American countries. Control in the sense that the state did not stop at the initial importation of the production machineries, but ensured that the local capacity to replicate most of their spares was developed. For instance, the choice of foreign technology that came into Korea through the importation of heavy machineries and other capital goods, was monitored by the state-owned agencies. Not only that. As part of its industrial policy, the state had, through its incentives like interest-free loans and tax holiday, encouraged specific local businessmen to establish trading companies with the hope of taking full charge of the implementation of the import substitution strategy as contained in then General Park regime's First Five-Year Economic Development Plan, 1962-1966.

As noted earlier, the trading groups were largely a creation of the Korean state. Hence, they were so intricately financially tied to the government so much so that they could hardly be regarded as private companies even though they pretended to be. As the government's vehicles for rapid economic growth and export-oriented industrialisation, the big business groups were involved in virtually all sectors of the Korean economy. In fact, they were a microcosm of the state's mode of extracting economic surplus. For instance, while the state provided the general economic and industrial policy guidelines on import substitution, the task of implementing them was undertaken by the big business groups. All this made it pretty difficult for foreign capital to be directly involved in any sector of the Korean economy throughout the 1962-1966 plan period.[18]

By 1967, the import substitution industrialization strategy which the General Park regime introduced in his first plan period, 1962-1966, had yielded some significant positive results. For instance, a 25 percent growth rate was recorded in the textile sub-sector of Korea's manufacturing industry. At issue, however, was not so much with the growth in some sectors of the country's manufacturing industry, but its deepening technological dependence on the developed market economies. For instance, inspite of the commitment of the administration of then General Park to Korea's industrialization, most of the production machineries and their spares used in the country were still sourced through import. In essence, Korea's manufacturing industry was yet to transcend the limitations of import substitution as of 1967. And until this problem was overcome, it would be difficult for Korea to wrest control over technology from foreign capital even though it was not directly involved in the operations of the economy. Hence, rather than reduce, the growth recorded in Korea's economy took place amid the country's increased importation of intermediate and capital goods. In 1968, that is, three years after the first economic plan period had come on stream, Japanese private capital and technology still dominated Korea's economy since a lot of the capital goods were imported from Japan. All this eventually helped structure General Park administration's next economic development plan including industrialisation .

Therefore, the Second Five-Year Economic Development Plan of General Park's administration, 1968-1972, was launched to address among others, the deepening technological dependence and weak industrial base of South Korea. The second economic development plan was regarded by the Korean government as a period of 'deepening Korea's industrial base'[19], with particular emphasis on the development of the steel sector. Then Park regime was of the view that, its export-oriented industrialization would run into difficulties if the heavy and chemical industries were to be wholly dependent on imports for their production inputs and spares. It was in an attempt to resolve these and other contradictions that steel development was embarked upon in the second economic development plan. In essence, the Korea government only began to develop its steel sector in the the late 1968, about the same time that the Nigerian government also started its steel projects.

Though the development of the steel industry in both Nigeria and South Korea began about the same period and was state-led, their experiences were strikingly different. As noted earlier, steel development and industrialization among others, were no less than a political project and as an *alibi* used by the state managers and the political class to hang on to political power and to amass wealth, but not to really reverse the country's industrial backwardness. This was in sharp contrast to the approach the Korean state took to the development of its steel industry. In Korea, the state embarked on the steel project with all the seriousness and purposefulness that it required to succeed in helping bring about the industrialization of the Republic.

In 1968, the regime of General Park Chung-Hee set up a 'Committee on Steel Development' under the chairmanship of another Park, who was a retired army general and one of his close confidants. One of the hallmarks of then Park's administration was that, it used retired military colleagues in government so much so that it bred cronyism, a trend that also pervaded Nigerian government since 1966 when the military first seized power.

The danger of this approach to governance is that, public policies and projects have, in most cases, been geared towards the satisfaction of the parochial interests of a cabal in the military that is in control of political power, but not centered on the well-being of the majority of the people. This was quite evident in the approach that then General Park took to governance and economic development of Korea. Apart from the steel sector, his regime relied on both retired and serving military officers for the management of other public companies engaged in sectors like telecommunication, real estate, railways and power. They also dominated partisan politics in South Korea. Inspite of the assassination of General Park in 1979 through a palace coup which was reportedly led by then chief of Korean Central Intelligence Agency [KCIA], the trend continued into the 1980s. Cronyism was so widely spread across all facets of the Korean economy and society with the military men, both serving and retired, being the greatest beneficiaries that it posed a formidable obstacle to the enthronement of democracy in Korea. Inspite of two successive civilian governments after the overthrow of General Park in 1979, military cronyism still held sway in Korea's local politics and in its economic development plans. It was little wonder that the Korean government's first major steel plant, Pohang Iron and Steel Company, [POSCO] was located at Pohang, the home town of General Park who headed the Committee on Steel Development. The site of the company was a former rice field, though close to the sea. But if accessibilty was a factor, most of the Koreans the author spoke with in Seoul and Pusan were of the view that Pusan would have been a better choice since it has a deeper and narrower continental shelf than Pohang.

Generally, the majority of the Korean publics and scholars were of the view that it was not until the election of Kim Young-Sam as Korea's president in 1993, that Korea could be seen to be gradually moving towards democracy as the regime of then president Kim and his ruling party fought to break the stronghold which the retired generals had had on the polity. Not only that, previous regimes were seen only to have tried to mimic some of the traits of democracy whereas they were indeed, really dictatorial. In other words, attempts that have been made by the Korean political class and government particularly since 1993 when president Kim Young-Sam till the incumbent Kim Dea-Jung at the time of the study to democratize the polity and economy should serve as one important lesson to Nigeria's approach to politics and economic development. That is, it is important for a country's economic and industrial development including its steel project to be founded on democracy. Time and space will not allow us to elaborate more on this.

Inspite of the prevalence of cronyism in Korean politics, the steel project was still kept on course. The Park's Committee on Steel Development was charged with duties, among others, to carry out feasibility studies on the establishment of an integrated steel company in Korea, taking cognizance of the financial needs, raw material, the technological route to adopt and the product mix. It was also to formulate particular steel policy that will guide steel development in the country. In many respects, the mandate of the Committee was similar to that of its counterpart in Nigeria, the NSDA. For instance, with a poor industrial base at independence, both the Park Committee and the NSDA invited the major Western European steel producing countries to help them set up their steel companies. One of the major differences between them, however, was that Nigeria, had most of the steel raw materials which South Korea lacked; and this went a long way in influencing the outcome of their reports.

Given Korea's poverty in most of the basic steel raw materials notably iron ore and coal, the Park Committee had a greater task to accomplish in comparison with the NSDA in Nigeria in the sense that the latter had to recommend modalities for the external sourcing of steel raw materials. Among the major European manufacturers of steel plants that the members of the Committee held consultations with and it cooperated, was the Voest Alpine of Austria. Britain, France and Germany among other major steel producing European countries, declined to sell the kind of steel making and producing equipment that Korea wanted: integrated steel plants with facilities to roll various diameters of flat steels and round profiles. Japan also refused to assist Korea in having its steel plants equipped with similar production lines. With the exception of Voest Alpine of Austria, the question is: Why did Britain, France, Germany and Japan oppose the installation of flat steel components in Korea's steel companies? In addition to the initial plan of the major Western European steel producing countries to leave South Korea as a steel consuming nation just like as they wanted Nigeria to be, the real intention was leave South Korea's steel sector outward-oriented after coming on stream. Nigeria had also faced similar opposition, but unsurprisingly, bowed to the dictates of the West as the state managers were more concerned with what would accrue to them as kick backs than using steel to quicken the country's industrialization. Furthermore, Voest Alpine which also installed the concast plant of Nigeria's DSC, struck a deal with the South Korean government not really because Austria supported Korea's steel project. On the contrary, as one of the leading manufacturers of concast plant and their spares, what really was of interest to Voest Apline, was to sell its products to South Korea, and nothing more. This was all the moreso as the European steel market was already getting overconcentrated and posting declining returns on investment.

After having toured major European steel producing countries like Germany, France, Belgium and Britain; and having also held negotiations with the big steel companies in Europe and Japan, the Park Committee's recommendations included: [i] that an integrated steel company should be established with an initial installed annual capacity of 1.03 million tonnes at the first phase, which would be increased in the next three phases to a maximum of 45 million tonnes at the fourth stage; [ii] that the major technological route for making and producing steel should be the blast furnace, though others like direct reduction and COREX, could be used; and [iii] that the production lines should be equipped with facilities for the production of various steel products notably flat steel and long profiles.[20]

Because Korea's nascent capitalist class was basically mercantile and did not have both financial and technical capacites to venture into steel development as explained earlier, the Committee advised that it should be encouraged to participate in the downstream sector of the steel sector, where less capital outlay would be required. The Committee was also of the view that, restricting local private capital to the downstream sector would help nurture the local businessmen into an industrial domestic bourgoisie that would assist the state in consolidating Korea's industrial base. Though not strictly as a recommendation of the Committee, it was also of the view that, given Korea's poverty in steel raw materials and lack of investment capital, the upstream sector of the country's steel sector should be dominated if not monopolised by the state.[21] However, as it shall soon be made clear in the course of the discussion, the state did not have it as a policy to officially prohibit local private investors from engaging into the upstream sector of the country's steel industry. Instead, its position was that, private steel companies that ventured into the upstream sector of the country's steel industry would have done so at their own cost as they were unlikely to benefit from the state's incentives on the heavy and chemical industries like interest-free loans, tax holidays and special policies for export promotion.[22]

The government of General Park finally adopted the Park-led Committee Report with minor amendments. In April 1968, the Korean government incorporated the Pohang Iron and Steel Company Limited [POSCO] with the government as the sole shareholder while retaining the Committee's previous Chairman, General Park [retired], as the Company's chief executive. Civil works commenced in 1970 with the reparation paid by the Japanese government. POSCO finally came on stream in 1973, when then General Park regime's Third Five-Year Economic Development Plan was launched. The ownership structure of POSCO was similar to Nigeria's DSC and three steel rolling mills in the sense that, both had the state as the only shareholder. However, as POSCO expanded, some of the agencies of the Korean state like the Korean Development Bank was later invited to hold equity position in the company in order to make it resemble a joint-stock company, whereas it was nothing of the sort. The Nigerian government had tried to do the same when the state embraced the Bank/Fund-led adjustment package in 1986, but failed. More details on this later.

POSCO was planned to launch South Korea's industrial and technological revolution. As a result, the company played the role of a regulator for the country's iron and steel industry. Towards this end, POSCO had a near monopoly of the upstream sector of the country's steel industry. As the only and largest integrated steel company in the Republic of Korea, POSCO's wide range of products were meant to help establish companies in the downstream sector of the country's steel industry. It was also part of POSCO's duty to use its production cost profile as a reference for other steel companies operating in the country. POSCO monitored the volume and type of steel products that were imported into the country. All this made the company to act not only as a 'gatekeeper' for the development of Korea's steel sector, but also, the government's sole monitor of the global steel industry. Long after the overthrow of the Park regime in 1979, subsequent governments in Korea still have not really deviated from the thrust of his industrial policy on the heavy industries.

It was with the coming on stream of POSCO that Korea's steel industry began to take shape. Furthermore, its establishment gave rise to other steel companies. For, apart from Inchon and Union Steel Companies that pre-dated POSCO, virtually all the private steel companies numbering over 80 and operating in the downstream section, were established mostly in the 1980s when it [POSCO] had begun full production. What is more, they were largely dependent on POSCO for most of their basic production input notably finished steel products like flat steel, round and long profiles, and semi-finished ones such as steel billets. Tables 3.3 and 3.4 have aptly summed up the structure of the Korean steel industry. Locate Tables 3.3 and 3.4 in the Statistical Appendix.

3.4 CONCLUDING REMARK

One of the hallmarks of the state's efforts at steel development in both Nigeria and South Korea, was that, it was state-led. While in Korea, it was the policy of the state to be at the heart of the steel sector in an attempt to order the country's industrial and technological growth, the reverse was the case in Nigeria. Whereas POSCO had largely structured the direction which the private steel companies in Korea took, in Nigeria, the public integrated steel company, that is, DSC took off on a wrong footing and therefore, unable to play a similar role like Korea's POSCO. Hence, it was out of phase with the development in the steel industry. What is the structure of the steel industries of Nigeria and South Korea ?

NOTES AND REFERENCES

1. See Daniel Omoweh,'The Nigerian Iron and Steel Industry and Adjustment' in Adebayo Olukoshi ed., *Crisis and Adjustment in the Nigerian Economy*, [Lagos, JAD Publishers, 1991]; and his 'Nigeria's Steel Sector in the Global Steel Industry', *Annals, op. cit.*
2. ibid.
3. For details, see Carter J. Eckert et al, *Korea Old and New: A History*, op. cit.
4. See Ducan Burn, *The Steel Industry, 1939-1959: A Study in Cooperation and Planning*, [Cambridge, Cambridge University Press, 1961]; Louis Lister, *Europe's Coal and Steel Community: An Experiment in Economic Union*, [New York, Twentieth Century Fund, 1960]; The OECD, *The Iron and Steel Industry*, 1975. For more details, see also, W. Johnson, *The Steel Industry of India*, [Mass., Harvard University Press, 1966].
5. See O. Ekundare, *Economic History of Nigeria... op. cit.*; and Peter Kilby, *Industrialization in an Open Economy... op. cit.*
6. Peter Kilby, *op. cit.*
7. Carter J. Eckert *et al op. cit;* and for details on Nigeria, see, O. Ekundare, *op. cit.*
8. Carter J. Eckert *et al ibid.*
9. *ibid.*
10. See O. Ekundare, *Economic History of Nigeria... op. cit.*
11. Carter J. Eckert *et al.*
12. *ibid.*
13. *ibid.*
14. See Daniel Omoweh, *op. cit*; and Peter Kilby, *Industrialization in an Open Economy... op. cit.*
15. As the minister of Mines, Power and Steel during the second republic, 1979-1983, the late Malam Ali Makele was one of the wanted persons by then Buhari-led military regime to account for the huge fund spent on the Nigerian public steel companies where not much work was done.
16. Based on the author's field trips in South Korea.
17. See Kyan-Hwie Mihn, *Industrial Policy for Industrialization of Korea*,

[Seoul, KIET, 1988]; and Jong-Chan Rhee, *The State and Industry in South Korea: The Limits of the Authoritarian State*, [London, Routlegde, 1994].
18. *ibid.* See also, Alice Amsden, *Asia's Next Giant... op. cit.*
19. This was also the official view of the Korean Development Institute, KDI. The KDI was one of the most respected policy-making organs of the Korean government at the time this study was conducted.
20. See the General Park-led 'Committee on Steel Development Report' submitted to the Korean government, 1969. Additional data were obtained by the author during interviews with the officials of POSCO in Seoul and Pohang.
21. *ibid.*
22. At one of the roundtables with the author in Seoul, officials of POSRI partly blamed the bankruptcy of Hanbo Steel Company for failure to operate in the downstream sector of the country's steel where it would have benefitted from some of the state's incentives. However, the officials of Korea Steel Association told the author that even some of the steel companies that operated in the downstream sector still did not benefit from the state incentives as they remained in protracted financial crisis as of September 1997.

Chapter Four

Structure of the Steel Industry

4.1 Introduction

This chapter critically examines the structure and operations of the steel industries of Nigeria in comparison with South Korea's, taking cognizance of their product mix and linkages with the intermediate and capital goods-producing companies in particular, the larger economy generally. It pays special attention to the structure of the steel industries of both countries since it gives useful insight into the extent of the countries' level of industrialization.

4.2 Structure and Operations of the Steel Industry

Structurally, the steel industries of both Nigeria and South Korea, like those of the economically advanced countries of Japan, Germany, Britain and France, can be generally divided into two broad sectors: upstream and downstream. The companies involved in the upstream sector of the steel industry are basically engaged in the exploration and exploitation of the basic steel raw materials including the alloying minerals, making iron and producing steel, and producing semi-finished and finished steel products. In the downstream sector on the other hand, the companies are engaged in finish-up operations. They use either the semi-finished or finished products, or even both products of the companies that operate in the upstream sector to produce intermediate and capital goods. Let us elaborate more on the structure of the steel sectors of both Nigeria and South Korea, beginning with the upstream sector.

4.2.1 UPSTREAM SECTOR

4.2.1.1 Nigeria

To begin with, until the state intervened in steel development process in the late 1970s, there was really no steel company operating in the upstream sector of Nigeria's steel industry. Even long after the submission of the reports of the country's geological and geophysical mappings to the Nigerian government by the Russian experts which favoured the setting up of steel companies in the country, it was still difficult embarking on steel development due largely to lack of finance. This perhaps partly accounted for why the exploitation of coal since the period of British colonial rule, was directed for export. Worst still, the establishment of two integrated steel plants one of which uses coal, did not force the government into rethinking how the local coal deposit could be used locally. Inspite of the reports from the SRMEA which showed mineable reserves of alloying minerals like tantalite and vanadium, they were still not developed largely as a result of the indifference of the state to steel development and industrialization. What then, is the structure of the upstream sector of Nigeria's steel industry?

Led by the state, the upstream sector of Nigeria's iron and steel industry was still at a fledgling one. Ninety-eight percent dominated by the state and its agencies, the remainder was accounted for by the private steel mini-mills at the time this study was undertaken. The activities of the state in the upstream sector consisted largely of the establishment of integrated steel companies, the exploration and exploitation of iron ore, and the provision of energy in the forms of electricity and natural gas to fire the steel plants. As noted, though efforts were made by the state to explore for local coal deposit that could blend with imported coal for use at the Ajaokuta steel plant, coal mining *per se* could hardly be regarded as part of the activity in the upstream sector of the country's steel industry. This is because coal, like petroleum, is primarily mined for export. The same can also be said of the gas industry where the state and the operating foreign oil companies have been flaring virtually all the gas produced since they began oil exploitation over four decades ago. In essence, our discussion will be limited to the integrated steel plants, the mini-steel mills, and the exploration and exploitation of steel raw materials, though mention will be made of the coal and gas industry to clarify or enrich our argument.

The two integrated steel companies, namely, Delta Steel Company and Ajaokuta Steel Company [still a project]; and the National Iron Ore Mining Company, Itakpe, were all owned by the state. Other government-owned agencies involved in the upstream sector of Nigeria's steel industry included the NMDC, Jos; and the SRMEA, Kaduna.

As an integrated steel company, DSC was designed with direct reduction technological route for first converting iron ore into iron and then steel, which, in turn, is used to manufacture both finished and semi-finished steel products. Locate Figure 4.1 in Statistical Appendix which summarises the company's production flow chart of iron and steel making. The company was also equipped with a captive foundry and a melting shop to forge most of the basic spare parts like bolts and nuts for in-house routine maintenance of some of its machineries. The company's in-house rolling mill was planned to use part of the billets cast by the company to roll out steel rods and coils. DSC's overall product profile consisted of steel billets, steel rods, coils and bars. Although its foundry could fabricate some intermediate and capital goods like crawsher balls, bolts and nuts as well as reground cam and crank shafts among others, they were mainly for internal use and not really as part of its product profile. DSC was not fitted with components for rolling out flat steel sheets due in part to how the state managers have conceived industrialization. And if eventually Ajaokuta comes on stream, its product range would not be significantly different from that of DSC in the sense that there was still no plan to incorporate into the company's production lines, facilities for producing flat steel at the time this study was conducted. Unlike DSC's direct reduction technologically process of making iron and steel, that of Ajaokuta Steel Company was based on blast furnace. Locate Figure 4.2 in Statistical Appendix for details on the production flow of iron and steel making through the blast furnace.

While DSC was built on turn-key by a group of Western European steel companies notably, Voest Alpine of Austria and various steel companies from Germany, the construction of the Ajaokuta's blast furnace by the TPE of Russia, was mechanical. That is, it is not based on turn-key. The direct reduction method is a much more recent technology compared with the traditional blast furnace. Even then, both technologies are still used for making steel by the advanced steel-producing countries till date. With respect to the private mini-steel mills, they were built by the European transnationals like Ferrostaal. They were planned to recycle scrap iron and steel to cast steel billets rolling internally. They also relied on imported steel billets to roll out steel rods and coils.[1] Figure 4.3, located in Statistical Appendix, summarises the operations of a mini-steel mill. For the purpose of cost reduction and more effective operation, rather than build large integrated steel plants, mini-mills are gradually been transformed into integrated steel companies of modest sizes by incorporating facilities for iron and steel making as well as production lines for rolling flat steel product. The experience of South Korea is quite instructive in this regard.

With the various installed capacities of DSC and Ajaokuta, and the few private mini-steel mills combined, it was expected that it would be possible to produce about 5 million metric tons of crude steel annually in Nigeria if all went well with steel development. Of this, DSC was expected to produce 1.5 million metric tons, representing 30 percent of the total annual output of crude steel produced. Out of the remaining 70 percent, while the private mini-steel mills would produce 0.5 million metric translating crude steel, the Ajaokuta Steel Company, if it finally comes on stream, would produce the lion-share of 3 million metric tons, representing 60 percent of the country's expected annual output of crude steel.[2]

DSC was installed with equipment that could cast 960,000 metric tons of steel billets out of its total crude steel produced. Out of this, 360,000 metric tons of steel billets representing 37.5 percent, were to be supplied to the government-owned steel rolling mills located at Katsina, Jos and Oshogbo where they would be rolled into steel rods and coils of small dimensions. The remaining 62.5 percent, amounting to 600,000 metric tons of billets, would be used by the company's in-house steel rolling mill. At the completion of Ajaokuta, it was expected that 2.5 million tons of billets would be cast by the company yearly with all of them to be consumed by its own steel rolling mill. About 250,000 metric tons of steel billets were planned to be cast by the private steel mini-mills, bringing the total to 3.71 million tons of billets annually. Also, the in-house rolling mills of both DSC and Ajaokuta had a total installed capacities to produce 2million metric tonnes of rolled products at their fourth and final phases, though DSC that was completed and began operation since 1982 was still at its first phase as of 1998. See Table 4.1 for the product mix of DSC located in the Statistical Appendix. The remaining three public rolling mills located at Katsina, Jos and Oshogbo had installed capacities to produce additional 1.2million metric tonnes of rolled products in their final phases as well. The private steel rolling companies also had capacities to produce 500,000 metric tonnes of rolled products. Together, the steel rolling mills had a potential capacity to produce about 3.7 million metric tonnes of rolled products which consisted mainly of steel rods and coils.[3] All this does not mean that the public steel companies were planned to complement those in the private steel sector. Rather, they operated independent of each other. The lack of interlinks between the public and private steel companies stemmed in part from the absence of both industrial and steel policies.

Though the state seemed to be serious with the development of its steel industry particularly in the context of local sourcing of steel raw materials and setting up of other support services, its concern was basically to meet the needs of only the government-owned steel companies. Why? The thinking of the state was that, since it was only the public integrated steel company that operates in the country, it was of no use incorporating part of the activities of the private steel companies into its own steel development agenda. Whether it is wise to develop a country's steel industry along the thinking of the managers of the Nigerian state will be taken up in the course of the discussion. Even on their part, the private steel companies were not prepared to cooperate with the public steel companies given the nature of the foreign investors. Hence, they took off with the plan to source for their own raw materials both locally and externally; but not to rely on the state. Let us dwell more on the approach that the state took to the development of steel raw materials for the country's steel industry.

With respect to iron ore, the SRMEA had, carrying out preliminary exploration, put its local reserves at 1.5 billion metric tons as of 1988. No new exploration for iron ore has been made since then. Of this deposit, it was not until 1986, that is, four years after DSC had come on stream, that skeletal mining only began at the Itakpe deposit by the state-owned National Iron Ore Mining Company, NIOMCO. The Itakpe deposit had an estimated deposit of 300 million metric tons and of ferrous grade of 38 percent. The Itakpe iron ore has to be upgraded to 64 perecnt ferrous content so that it can be used at the Ajaokuta Steel Company. With further upgrading of the Itakpe iron ore to 68 ferrous content, it can be used by DSC.[4]

Like the SRMEA and NMDC which were created out of the defunct NSDA and later became autonomus government agencies, the NIOMCO was transformed into a parastatal. Its primary duty was to produce all known deposits of iron ore in the country, though the company's activities were still limited to the development of the Itakpe iron ore at the time this study was conducted. According to the NMDC, the Agbaja iron ore deposit with a ferrous grade of 67 percent, has an estimated proven reserve of 1 billion metric tonnes, the largest discovery ever made in the country. The high ferrous content of the Agbaja ore made it suitable for direct use by DSC, after de-sulphurising it. Which means that, if all had gone well, DSC would have come on stream with 100 percent locally produced iron ore. Even as of 1998, the government still did not make attempt to develop the Agbaja reserve.

As for limestone, another major raw material for iron and steel making, Nigeria has huge reserve. DSC came on stream wholly dependent on limestone produced by the Cross River Limestone Company, a state government-owned company exploiting the Mfamosing's 30 million metric tonnes deposit. But the same government decided to import the limestone that was used for lining the walls of the coke oven battery of the Ajaokuta steel company rather than using the locally produced limestone. This is one of the contradictions inherent in the approach that the state took to industrialization. That is not all, as reports from the SRMEA even showed that about 500 million metric tons of limestone that abound locally, are yet to be developed.[5]

Exploratory reports from the SRMEA and the Nigerian Coal Corporation estimated Nigeria's coal reserve at 500 million metric tons as of 1995. The NMDC's laboratory analysis of the samples of the Obi/Lafia coal deposit located in the Benue trough estimated at 162 million metric tonnes out of the country's total undeveloped coal reserve, showed that it had high metallurgical value and was cokable on its own, but of a higher cokable value if blended with imported coal preferably from Australia. This means that the Nigerian coal, contrary to the politicised report of its low metallurgical value, can indeed, be used for making iron and steel. Yet, when the coke oven battery at the Ajaokuta steel company was built, the government did not explore the option of using it. Instead, the coke oven was lit in 1989 with imported coke, which is a semi-processed coal with higher metallergical value. See Table 4.2 in the Statistical Appendix for the state of Nigeria's steel raw materials.

About three percent out of the 8 trillion cubic meters of the country's natural gas being wasted through flaring annually, was utilised in the economy. Of this figure, the steel companies notably, DSC consumed about 30 percent; and an additional 40 percent to be used by Ajaokuta steel company when it becomes operational, and the remainder to be used by the government for generating electricity. Both DSC and Ajaokuta were planned to be 100 percent dependent on natural gas to fire their steel plants. Together, DSC and Ajaokuta, with the three inland steel rolling companies, were however, to rely on the state-owned National Eletric Power Authority, NEPA, for other power supply than for the purpose of firing the steel plants. Yet, with the singular exception of Jos Steel Rolling Company, each of the remaining public steel companies had a captive power plant due in part to the protracted inability of NEPA to supply power to them. As for the private steel companies, they had their own captive power plants though, they still used the power supplied by NEPA whenever it came.

There was incoherence in the path that the state took to the provision of electricity for not only the steel companies, but also, to the manufacturing sector and the country at large. It is a contradiction that stems in part from the state's indifference to energy development. Without an energy policy for instance, each of the public steel companies has, rather than be part and parcel of an integrated energy structure, wanted its own captive power plant. The same was true of the private steel compamies. The overall consequence of all this, is the incessant power outage which has crippled the operations of the steel companies before they were shut down indefinitely.

Generally, the technologies of iron and steel making are to large extent, universal except for some minor adjustments that are made to adapt them to recepient countries' needs. What it all means is that, ideally, the steel technologies that were employed in Nigeria ought to be adapted to enhance the nation's technological capability as well as use its steel raw materials among industrial minerals that abound locally. As part of the effort to deepen the country's technological base for instance, the steel companies should be made to produce flat steel and and establish foundries that could cast locally, most of the major intermediate and capital goods previously imported. None of this hardly entered into the calculations of the state when it embarked on steel development for reasons already explained. As a result, DSC and the inland rolling mills located at Jos and Oshogbo were built on turn-key and were fitted with computerized production systems. The implications of this are many and grave. At DSC, for instance, it all means that, right from the stage of iron and steel making through the casting of billet and then loaded into the furnace for charging before being rolled into produts, and finally, the finished products were packaged for sale, the processes involved were all computerized. The same is true of the steel rolling companies, the only difference between them and the integrated steel plants is that, the former was not involved in iron and steel making.

Furthermore, DSC had no capacity to replicate the major spares of its major production machineries like the iron and steel making plants. What is more, the Ajaokuta Steel Company, which if eventually completed, will be based on another technology for iron and steel making, that is, the basic oxygen furnace. Like DSC, the Ajaokuta was not fitted with facilities to cast the spares of its major plants. Within the public steel companies, therefore, there were two basic steel technologies: one modern [DSC]; and the other traditional [Ajaokuta]. At issue is not so much with the steel technologies as the lack of internal technical capacity to cope with their maintenace. The same technological dependence pervaded the steel rolling mills at Jos and Oshogbo which were operated on automation. It was only the Katsina steel rolling mill that was built on non-automated system by the Kobe Steel of Japan. That is to say, it is operated manually. Even then, it has no internal capacity to indigenize the technology for rolling steel by way of duplicating parts of the furnace, nor any other components of the rolling mill.

The situation in the private steel companies was not any different. Given the facilities of the mini-mill, notably, CISCO and Universal Steel, all that they were able to do so far, was to melt scrap to make iron and steel, and cast billets for internal use. However, this was irregular since they lacked the funds to procure production inputs. In essence, though the mini-mills represented a third technological route in the upstream sector of the country's steel industry, they were dependent on the countries of the North for their technological needs. Worst still, their activities were guided in secrecy. Though they were members of the Manufacturers' Association of Nigeria, MAN, the Association did not have any useful information on them, particulary their equity structure, technology and sources of raw materials. And as a consequence, there was little or no information on their interlinks with the government-owned steel companies, and in any case, there were cogent reasons for doubting if there were any.[7]

From all appearances, the upstream sector of Nigeria's steel industry was disarticulated from the onset. Hence, it was heavily dependent on the Western steel transnationals. What is the situation in the upstream sector of South Korea's steel industry ?

4.2.1.2 *South Korea*

The structure and operations of the upstream sector of South Korea's steel industry contrasts sharply with Nigeria's in many respects due in part to the differences in their endowment with natural resources for the steel industry. For instance, while Nigeria has large deposits of high grade iron ore and proven reserves of metallurgical coal with which it could have started its steel sector relatively well but failed to do so, the opposite is the case for

South Korea as its upstream steel sector is virtually import-dependent.

In a bid by the Korean government to actualise its policy of 'deepening industrialization', the steel sector was accorded a high premium in the country's agenda of development. This was all the more so since the regime of General Park was of the view that, it would be impossible to implement the government's industrialization programmes of developing the heavy and chemical industries without first consolidating the steel sector. As explained earlier, given the poverty of South Korea in steel raw materials and funds, the thinking of the the regime was that, the development of the upstream sector of the country's steel industry should be solely handled by the state. Though it was not really part of the general industrial/steel policy of the Korean state to monopolise the upstream sector of Korea's steel industry, it turned out to be so in practise. It was in this context that the Korean government established Pohang Iron and Steel Company, POSCO.

As part of its duty, POSCO was to chart the course of Korea's overall steel industry in particular and its industrial and technological growth generally. Within this context, POSCO did a lot to restructure the country's overall steel industry in compliance with the goal of its national industrial policy. For instance, POSCO dominated both the up- and downstream sectors of Korea's steel industry, controlling 95 percent of the activities in the industry, while the remaining five percent was accounted for by the local private steel companies. This was almost similar to the situation in Nigeria only in terms of the percentage share of the public steel companies of the overall activities in the country's steel industry as they varied in their actual performances.

POSCO had two integrated steel plants: the first was located at Keodong-dong, Pohang which doubled as the company's corporate headquaters; and the second steel plant was sited at Kwangyang. The two steel plants at Keodong-dong and Kwangyang had a combined annual installed capacity to produce 38.90 million metric tons of crude steel as of December 1997. This was about eight times larger than Nigeria's. Both plants of POSCO were planned to be 100 percent imported-dependent on iron ore and coal since the Republic lacked these basic steel raw materials. Table 4.3, located in the Statistical Appendix, summarises the outlook of the import of iron ore and coal by POSCO. However, in terms of limestone which the country is richly endowed with, the steel companies were compelled by policy to source it locally,[8] this is similar to the situation in Nigeria. POSCO's main production facilities included: iron and steel making, rolling flat and round profiles, and producing galvanized and stainless steels, and manufacturing of production machineries and their spares. This is well documented in Table 4.4 located in the Statistical Appendix. Compared with Nigeria's DSC that could only produce basic steel rods and coils, this is another major contrast. As noted, even when the Ajaokuta Steel Company eventually comes on stream, it will still roll out the same kind of steel product. The question that arises is: What purpose does it serve Nigeria for having both DSC and Ajaokuta with a combined annual capacity to produce 4 million metric tons of crude steel from which only simple constructional steels like rods and coils were rolled? Though this might present itself as one of the major policy errors on the part of the government officials who, as argued earlier, were entrusted with the physical development of the public steel companies, its origin resided with the managers of the state who never had industrialization on their plan. More details on this later.

With respect to its major products, POSCO had a wider range in comparison to its counterparts in Nigeria. Notable among POSCO's major products were: wire rods, hot and cold rolled coils and sheets, pickled and oiled coils, thin plate and coils, galvanized coils and sheets, electric steel sheets and strips, stainless hot and cold rolled coils and sheets, and thin organic coated steel sheets as already shown in Table 3.3 in chapter Three. This was in sharp contrast with the mono-product mix of Nigeria's steel companies including the steel rolling mills. One thing is clear from the nature of product of steel companies. That is, the product mix gives an insight into the approach the state has taken to industrialization. For instance, it was the industrial policy of the Korean state that largely guided the products of Korea's steel companies in the sense that they were tailored towards the achievement of the country's industrialization. And in Nigeria where there was yet to be an industrial policy as of 1998, it was little wonder that the product mix of the steel companies was still biased for simple construction as opposed to industrialization.

In no other area has the difference between the upstream sector of the steel industries of Korea and Nigeria been more outstanding than in the sourcing of steel raw materials. While the upstream sector of South Korea's steel industry was about 100 percent dependent on external sources for its basic steel raw materials, Nigeria's, which ought to be self-reliant in steel raw materials since they abound locally, has been largely import-dependent for reasons already explaind. Since South Korea lacked iron ore and cokable coal, many a policy maker, would ordinarily have considered it economically unwise to embark on a steel project in the first place. Korea's poverty would have even made the Government uninterested in steel development given the huge capital outlay required. Yet, the country's poverty in relation to Nigeria, did not really constitute any hinderance to steel development. Why was the state resolute in its determination to have a viable steel industry underway? It is partly because the Korean state came to realise that, the country could hardly become industrialized without having its steel sector put on a sound footing. This is all the more so since it is a 'late starter' in industrialization. In the light of all this, the sourcing of raw materials for Korea's steel sector ranked very high on the state's agenda of industrialization programmes. To ensure a steady source of raw materials, the Korean state adopted a two-pronged approach: First, it invested in the joint development of reserves of iron ore and coal in overseas notably Australia and Canada. Second, it imported steel raw materials directly from the producing countries. South Korea's major source of iron ore was Australia, with Brazil and India ranking second and third respectively. While the Australian iron ore deposit was developed jointly with POSCO, the company bought iron ore from Brazil and India, and at times, from Peru. As for coking coal, POSCO imported directly from Australia and Canada. The company was also dependent on import for alloying minerals like vanadium, tantalite, and chromium.[9]

In fact, POSCO's overall import dependency ratio of basic steel raw materials and alloying minerals was 99.9 percent at the time this study was conducted, and there were indications that it would remain so for a long time to come. Out of the company's total imports of steel raw materials estimated at 56,533,000 metric tons in 1995 for instance, iron ore accounted for 59 percent which translated into 33,184,000 tons, while coking coal took 15, 953,000 tons representing 28 percent. Other imported steel raw materials outside of the joint venture arrangements, were metal scraps which accounted for 9 percent of the company's total imports, and the remaining 4 percent went to pig iron. See Table 4.3 located in the Statistical Appendix for details.

From all indications, South Korea's steel industry would continue to be dependent on imports for its major steel raw material. This is markedly different from Nigeria where the importation of iron ore and coking coal for use by the public steel companies could be stopped if only the basic fundamentals for steel development are got right and with a conducive political environment that will permit industrialization to take place. This is treated in greater details in the course of the discussion.

On technology, POSCO was equipped with the conventional blast furnace, the COREX method, and the mini-mill production processes for making iron and steel. The production process of POSCO's blast furnace was not significantly different from Nigeria's uncompleted Ajaokuta steel company in terms of design, except in operations where the former is fully automated and the latter was not. The details of the blast furnace technological route were already explained in Figure 4.2. South Korea's POSCO also had the COREX technology which is more modern than the Midrex technological route of Nigeria's DSC. In fact, it was only ISCOR of South Africa that operated the COREX technology in Africa as of 1998. See Figure 4.4 located in the Statistical Appendix for details.

The company also operated an integrated mini-mill which could serve as the third technological route for iron and steel making in South Korea. Its mini-mills were planned like the major integrated steel plants except that the moudles were of smaller capacities. Locate Figure 4.5 in the Statistical Appendix for details on its production flow.

About 98 percent of South Korea's annual total crude steel of about 40 million metric tons was produced through the blast furnace. The COREX plant at Pohang Steel Works was only completed in November 1995. It accounted for about two percent since it was still being experimented with. There were no indications of its wider application in South Korea at the time this study was conducted.[10] Both the Midrex and the COREX methods were an improvement over the blast furnace in the sense that they do not require sintering and coking plants in the production of pig iron. However, while the COREX method by-passes natural gas in the production of pig iron, the Midrex method does not. According to the officials of POSCO that the author spoke to, one of the major reasons why the conventional blast furnace and electric arc furance still remained the dominant steel technology in Korea was because the country had no natural gas and could not afford to produce pig iron on large scale through the COREX approach due to the higher power input required to fire the plant. POSCO had to import pig iron to operate its COREX plant. In essence, even the COREX is unlikely to gain a wider application in South Korea much more contemplating introducing the direct reduction technological route.

On account of cost, POSCO had, after completing the fourth and final phase of Kwangyang Steel Works in October 1992, decided to use integrated mini-mills to expand its operations in the future. In addition to cost reduction, the integrated steel mini-mill requires less than a year to install compared to a minimum of three years used to build an integrated steel plant. It even operates both the blast furnace and electric arc furnace. It is also possible to have all the production processes of an integrated mini-mill, that is, right from iron and steel making through the rolling mills to the final stages where the finished products were ready for sale, fully automated in order to gain man-hour input.

Unlike in Nigeria where the private steel companies still operated independent of the public steel companies, in South Korea, they were significantly integrated into the activities of POSCO at the time this study was conducted. Until it became bankrupt in March 1977, it was only the Hanbo Steel Company among the private steel companies that operated in the upstream sector of Korea's steel industry. As noted earlier, POSCO dominated and controlled the upstream sector of the Korea's steel industry, a development that was vehemently opposed by the private sector. According to some of the officials of the Korea Steel Association, KOSA, that the author spoke to in Seoul in June 1997, it was all plots of the Korean state to own and control the real sector of the Korean economy. In an attempt to reverse the situation, the Association said that, although it had repeatedly made collective representations to the Korean government in the past in which it asked for special incentives such as financial assistance that would enable the member-companies intervene in the upstream sector of the industry, no positive result was achieved. The pertinent question is: Why was the Korean state reluctant to assist the private steel companies to come into the upstream sector of the industry? Contrary to the position of the Korean government that the high import dependency ratio in the country's upstream sector compelled it to limit the activities of the private steel companies to its downstream, the officials of the KOSA said that, it was because the government did not want them to operate independent of POSCO which acts as its 'gate-keeper' of the country's steel industry.[11] Compared with Nigeria, there was hardly an all-embracing association of private steel companies like KOSA, that served as the umbrella body for the private steel companies to meet regularly to discuss the problems and prospects of Korea's private steel sector. Rather, what existed in Nigeria, was the Manufacturers' Association of Nigeria, MAN, of which all manufacturing concerns including those in the steel sector are members. Unfortunately, the Association only had sketchy information on the public steel companies like DSC and hardly any on the private steel companies unlike in South Korea, where the KOSA kept complete and update data of its members, and has power to sanction any erring member- steel company.

By and large, the upstream sectors of the steel industries of both Nigeria and South Korea were at different levels of development. One of the major differences between them is that, while Korea had and would continue to have a high import dependency ratio of steel raw materials, in Nigeria, the current importation of basic steel raw materials would stop once the right fundamentals and leadership are in place. The upstream sector of steel industry is as important as its downstream since the level of development of both sectors would give a deeper understanding of the status of their industrialization. It is in order to discuss the downstream sector.

4.2.2 Downstream Sector

To begin with, the majority of the products of the steel rolling mills must of necessity, have to undergo further processing such as fabrication among others, before they are usable. It is with such additional processes that the technology for the production of both intermediate and capital goods really resides. Suffice it to note that, though the product mix of steel companies does give an insight into the technological capacity of a country, it is, indeed, the use into which the steel products are finally put largely by the industries in its downstream sector that really determines the level of the local technological capacity and industrialization of a country. The major aim of this section, therefore, is to describe as well as critically analyse the structure of the downstream steel sectors of both Nigeria and South Korea. It also serves as a background context for the analysis of the performance of the steel industries of both countries in Chapter Five. Let us begin the discussion with the structure of the operating companies in the downstream sector of Nigeria's steel industry.

4.2.2.1 Nigeria

In this section, One attempts a sketch of the structure of the downstream sector of Nigeria's steel industry, taking cognizance of the period of British colonial rule and how it helped shape its contemporary outlook, and its role in the country's industrialisation in terms of linkages with other sectors of the economy, especially the heavy industries.

As noted, the British had no plan to industrialise Nigeria during the colonial period, 1900-1960. And as a consequence, there was really no industrial policy that guided later, the establishment of some intermediate and capital goods-producing companies in the country. What took place instead, was for the British to encourage companies of British origin to relocate some stages in their production lines, particularly those involved in end-product activities to Nigeria, ostensibly to reproduce locally some of the goods that were previously imported. In the process of relocating some of the companies to Nigeria especially as the British shifted emphasis from purely exchange to production, it was the production machineries that were basically moved to commence the local reproduction of some intermediate and capital goods hitherto imported into the country. There was really no attempt made by the companies to duplicate parts of the heavy machineries. Nor did the companies source their raw materials locally as they were dependent on imports for both raw materials and the spares for the production machineries. Whether in fabrication engineering or assembly, the companies had a 100 percent import dependency ratio. This was the context in which import substitution strategy was applied in Nigeria.

As a consequence, the majority of the companies that constituted the downstream sector of Nigeria's steel industry, namely, steel rolling companies, cottage forge shops and foundries, were not only of small equity base, but privately owned. Why? It was part of the delibrate policy of the British to limit the installed production capacities of these companies since they were meant to complement imported intermediate and capital goods from the United Kingdom. Suffice it to note that, they were meant to keep the metropolitan factories in production while reproducing locally, some of these previously imported goods.

Not surprisingly, there was no discernible structure of the downstream sector of Nigeria's steel industry during the period of British colonial rule. The majority of the companies that operated in this sector were therefore, stockists of finished steel products like rods, pipes and angles, steel frames for door and windows, nails and barbed wires, and wire nets and meshes most of which were meant for constructional purposes. The business of importing finished steel products was dominated by the subsidiaries of the UAC like John Holt. However, there were some local trading companies like the Ibadan-based Sanusi Brothers Limited that was first a distributor to the major steel importers, before it later engaged in direct importation of some these goods from the UK in the late 1950s. There were some cottage forge shops especially among the Akwa people in the present-day Anambra state of Nigeria, who used scraps of iron and steel to fabricate farming implements like cutlasses, and cooking utensils like pots. However, the small scale of production and poor quality product of these products did not pose any threat to the imported ones as state policies and actions were used to guarantee the local market for the foreign trading companies.

After Nigeria had attained political independence in 1960, there was still no significant structural change in the downstream operations of the country's steel industry. The reasons are not far to seek. Like its predecessor, those in charge of the Nigerian state were only interested in inheriting but not changing the colonial policies and structures of exploitation. So, their conception of industrialization was really no less than a continuation of the approach which the forces of colonial capitalism, notably the state and transnationals, had adopted in the Nigerian economy. That is, to set up companies that are at best, nothing less than a relocation of some of the end-product stages of the production lines of the metropolitan factories. It was particularly so since the majority of the policy makers, who trained in the UK, wanted Nigeria to follow the path that Britain took to industrialization, notwithstanding the differences in the historical experiences of both countries. It was one of the greatest policy blunders ever committed by the custodians of the Nigerian state which still bedevilled the country's efforts at industrialization including its steel sector, with little or no hope of reversing it in sight. In order to have a clearer picture of the inherent contradictions in the path that the state took to industrialization and the implications for the structure of the downstream sector of Nigeria's steel industry, let us periodize the discussion beginning with the period, 1960-1980.

Within the first decade of Nigeria's political independence, 1960-1970, steel development was not in the calculus of the state. The state managers felt that it was more important to consolidate the economy than embarking on steel projects. The low *per capita* steel consumption made steel development all the more a non-issue. Therefore, steel development was left in the hands of both local and private capitals. Notable among the private steel rolling companies were, the Niger Steel Company, Enugu; the Continental Iron and Steel Company, and Universal Steel Company, both based in Ikeja-Lagos. Both companies also pioneered the rolling of steel products in the post-colonial period as the rolling mill of Niger Steel Company collapsed during trial production. CISCO and USC, were planned to import a larger portion of their raw materials, notably, steel billets; and to supplement this with the recycling of scrap metals to cast billets. They were also expected to import and market finished steel products like rods, angles and bars. However, while CISCO's major product mix was iron rods and coils, USC concentrated more in the production of enamel wares from imported flat steel than any other steel products.

High import-dependency ratio was also a major feature of the structure of the light metal manufacturing and fabrication companies that were established by both local and foreign capitals in the period under review. The majority of the companies were owned by private foreign capital largely of British origin. Reliant on light and medium steel bars and flats, they fabricated some intermediate and capital goods ranging from door frames, hinges, bolts, to metal furniture, storage tanks, waste bins, moulds, and machine parts. In most cases, they forged these parts based on the specifications of customers; and at times, as an interim measure to sustain production pending the arrival of the 'genuine' imported spare part from Europe. In essence, they were not in operation to really replace the foreign companies due in part to the sub-standard quality of their products in comparison to the imported ones. Notable among them were, Pressed Metal Works Company Limited, Ijora-Lagos; Metal Furniture Nigeria Limited, Ikeja-Lagos; and Roadside Engineering & Foundry Limited, Ilupeju-Lagos.

One unfortunate trend that came to dominate the structure of the companies the majority of which could be regarded as the thrust of the downstream sector of Nigeria's steel industry, was that they pre-dated the operation of the companies engaged in the upstream sector. Rather than having the activities in the upstream sector dictate the pattern of companies in the downstream as was the case in most of the industrialised countries like Germany and France, the reverse was the case in Nigeria. It was so because the activities of the companies in the upstream sector of the British steel industry largely shaped the structure of the downstream sector of Nigeria's steel sector. They were able to do this through the relocation of some aspects of their end-product operations to Nigeria. The nature of those in charge of the Nigerian state, particularly their thinking on industrialisation and mode of extracting surplus from the economy made it all the more difficult to reverse.

Following the boom in the global oil market in 1973, Nigeria was awashed with enormous wealth from oil export. It was part of the wealth that the government used to build social infrastructural amenities like road and public building. There was also increase in real estate development following the creation of more states and local government areas, which led to the construction of more basic social amenities like roads and bridges, public buildings, stadia and hospitals. All this raised the *per capita* steel consumption from 20kg to 150 kg in the 1970s. At this rate of local steel consumptiom, it became glaring that the private steel companies could not cope as importation of finished steel products continued to increase. In the 1970s, too, more light metal-based companies were established partly in an attempt to meet the steel needs and increasing demand of both intermediate and capital goods as activities in the Nigerian economy continued to grow. Notable among them were Kew Metal Works Limited, Ikeja-Lagos; Modern Engineering Works Limited, Sango-Ota; Steel Works Limited, Oluyole-Ibadan; Chister Engineering Works Nigeria Limited, Oshodi-Lagos; and Baltic Engineering Group Limited, Benin-City. Yet, the shortfall persisted. Not only that. Structurally, they all suffered from high dependence on import for their basic raw materials which, in most cases, were light and medium steel bars and flats. Although they recycled metal scraps for production purposes, it accounted for less than three percent of their raw material needs.

On account of the approach that the Nigerian state took to industrialization, the car/truck assembly plants were established in the 1970s to be wholly dependent on imports for their production inputs. The car assemby plants, a major aspect of the downstream operations of Nigeria's steel sector, pre-dated steel development in the country. So, it was no surprise that the heavy industrial sector, fragile as it was at the time this study was conducted, was totally out of phase with the country's steel industry. Again, this did not really bother both those in charge of the state and policy makers. It is not that they were unaware of the implications of having the majority of the heavy industries come on stream totally dependent on imports for their raw materials and spares for the production machineries. Nor is it impossible to redress the non-linkage of the steel industry with the heavy industries with an appropriate industrial/steel policy put in place. On the contrary, they were better off with the material benefits accruing to them as members of the board of directors of the subsidiary companies of the transnational corporations operating in conjuction with both local public and private capitals. These companies included: Peugeot Automobile Nigeria

Limited, Kaduna; Volkswagen of Nigeria Limited, Ojo-Lagos; SCOASSEMBLY, Apapa-Lagos; Boulous Enterprises Limited, Ogba-Ikeja, Lagos; and the Anambra Motor Manufacturing Company, Emene-Enugu.

In terms of raw materials, the heavy industries most of which are in the automotive sub-sector, relied on imported special flat steel sheets, steel and aluminium flat sheets and bars to fabricate some of their products, notably, bolts and nuts, fuel tanks, pedal systems, clutch discs, radiators, batteries, brake pads and linings, and oil seals. Others, essentially those into the assembly of cars, trucks, and pick-up vans, among others, relied on imported completely knocked down parts to produce these goods. In the strict sense of manufacturing, which involves the process of using raw materials to produce the finished goods therefore, the heavy industries were far from being engaged in it. Did the situation change following the establishment of the state-owned integrated steel companies?

Not really. For one thing, the state, as stated earlier, started steel development without any industrial policy particularly on how the steel industry could be used to redress the imbalance in the structure of manufacturing that Nigeria had inherited from the British. Right from the onset, what interested those managing the state and their cronies, was to take full advantage of the steel project to amass wealth. Little or no attention was paid to the product mix of the steel companies which ought to provide the basis of their linkage with the rest of the economy. And as a consequence, all the state-owned steel companies were installed with facilities for rolling out only rods, coils and bars; products which have very limited applications in the manufacturing industry.

From all indications, the public steel companies were not established to really re-shape the structure and operations of the companies that constitute the downstream sector of Nigeria's steel industry in such a way it will reverse the country's industrial underdevelopment. The Machine Tools Limited [MTL], Oshogbo illustrates this contradiction very clearly.

Machine Tools Limited, Oshogbo, a fully government-owned company, was established to manufacture lathe machines and their parts in Nigeria. Given the role its counterparts are playing in the downstream sectors of the steel industries of South Korea,

India, Taiwan among other developing economies, the MTL was expected to manufacture late machines and other capital goods. The initial understanding of the state for establishing MTL was that, it would depend on the public steel companies especially DSC and Ajoakuta Steel Company for its major production inputs, notably flat steel sheets. But it was the same state that did not equip the integrated steel companies for that purpose. This was one of the major contradictions in national policy which is why the MTL could not effectively commence production as planned. And eventually when the company commenced production, it had to import the machine spares it was expected to manufacture for the purpose of understudying them. Worst still, MTL had to import the complete knocked down parts of the machines it had planned to produce, ostensibly to assemble them moreso as the steel companies could produce any of them. The irony of it all is that, a company that was planned to produce machines and their spares is itself, turned into a mini foundry and wholly import-dependent for its basic production inputs.[12]

The state had initially planned to set up a big foundry that would complement the activities of the integrated steel companies, the MTL, small forge shops, and other manufacturing companies, particularly in helping them to fabricate major parts of their production machineries. According to officials of the Steel Department of the Ministry of Power and Steel, Abuja, the proposed foundry was so bedevilled by intrigues and politics of its location by policy makers and the techno-bureaucratic class right from inception that it never went beyond the drawing board.[13] But given the politics of the MTL, had the foundry even come on stream, it would have done little or nothing to further the process of Nigeria's industrialisation. In essence, what has been in place in the absence of such a central foundry, were light metal fabricating companies like the Nigeria Foundries Limited and Roadside Engineering & Foundry Limited, both based in Ilupeju- Lagos; Anchor Products Limited, Ikeja-Lagos; and Pressed Metal Works Company Limited, Ijora-Lagos. The majority of the heavy industries had to set up their own captive foundries where they forged simple spares like bolts and nuts, regrind cam and crank shafts of machines, and fabricate fuel tanks, storage tanks and grain silos. So, too, were the integrated steel companies at Aladja and Ajaokuta. Although the foundries of DSC and Ajaokuta had fabricated parts on request and specification for PAN, Nigerian Railway Corporation among others, it was an exception to the rule. For, the steel companies decided to commercialise their foundries as the state subventions to them began to dwindle. By and large, there was really no established interlinks between these companies.

The state of the aluminium industry, a major aspect of the downsream sector of the steel industry, was not any better than what obtained in other industries described above. The companies were largely small in terms of equity, and their installed capacities did not exceed 30,000 metric tons annually. Like the steel companies, their product mix was essentially for construction purposes: frames for doors and windows, office partition, cladding and roofing. Until the government established the Aluminium Smelting Company [ALSC], at Ikot-Abasi, the industry was dominated by private foreign investors, the majority of whom came from Western Europe and Japan. Even with the entry of the ALSC, the aluminium industry is still controlled by the foreign companies. Indigenous participation in the industry is still insignificant; and in the few cases where there was, it was based on technical cooperation in which the foreign partners supplied both the technology and the raw materials. All that the Nigerian partners did, was to serve as local contacts for marketing the products. Most of the local businessmen involved in the industry were largely marketers of the finished products. This was due in part to the early foreign aluminium companies in the country, notably ALCAN, ALUMACO, FIRST ALUMINIUM and TOWER which were themselves, basically importers of finished alumunium products like cooking utensils and roofing sheets. With the promulgation of the indigenization decree in 1972 as amended in 1977 before it was repealed by the junta regime of late General Sani Abacha, most of them were compelled to have Nigerians on their board of directors. Some of them had started to reproduce locally, some of the hitherto imported goods to complement import. As companies operating in the end-product stage of the industry, the basic raw materials for the aluminium companies consisted of aluminium billets and flat sheets, which were imported into the country and later fabricated into finished intermediate and capital goods like frames for doors and windows, bolts and nuts, and roofing sheets. If it were all well with the steel companies, public and private, the linkage between them and aluminium industry would largely have been in the manufacturing of moulds, other production machineries and their spares for the latter. In essence, there was really no linkage between and among the companies operating in the downstream steel sector and those in the aluminium industry .

Another major group of companies that constitute part and parcel of the downstream sector of Nigeria's steel industry are the vehicle/truck assembly plants. The nature of their production inputs, which, in most cases, are basically special flat steel sheets, bars, beams and angles, make them one of the most important downstream companies. Since both the private and public steel companies have no components to roll out neither of these products, it is natural to expect the vehicle assembly plants to depend wholly on import for their raw materials. It is a sector that was heavily dominated by the subsidiaries of the European transnational corporations until the Indigenization decree of 1972 came into force. With the onset of the decree, the state acquired only participatory shares in the companies and still left foreign capital with the control of the technology and the supply of the basic raw materials. Peugeot Automobile Nigeria Limited, for instance, is 40 percent French and 60 percent Nigerian. Though the 60 percent Nigerian holding was in majority, it was only participatory. This is because the French investors not only supplied the completely knocked down parts, but also, controlled the technology for the assembly of peugeot cars. The same was true for SCOASSEMBLY, a division of SCOA, Nigeria; and the Anambra Motor Manufacturing Company [ANAMCO], Enugu.

By the 1980s, a small fraction of local private capital began to intervene into the heavy industrial sector. It was more of a response by a small fraction of private local capital to its neglect and lack of support from the government over the years. Relying on capital saved from petty businesses conducted in the past inaddition to small loans from the banks, various groups of local investors signed technical agreements with foreign companies most of which were from Taiwan, Singapore and South Korea. Most of the companies were located around Nnewi in the present-day Anambra state, while others were established in Lagos,

Ibadan, Port-Harcourt and Sango-Ota in Ogun state. They produce some of the auto-parts for the automobile industry. The Ibeto Group of Companies is among the notable ones in the Nnewi industrial zone, producing batteries, oil seals on specification for PAN. Fabinna Nigeria Limited, Port-Harcourt; Ferodo Nigeria Limited, among other companies, also produced brake shoes, pads and linings from scraps. Most of these companies still import intermediate and capital goods to complement the ones they reproduce locally. Like the heavy industries, which they were trying to complement their activities, there was really no clear-cut linkage between them and the steel companies.[14]

Although the products of these companies could not compete with the imported ones, they were gradually beginning to gain some measure of acceptability not only within the Nigerian market but across the West African sub-region. In essence, with the right micro and macro-economic policy put in place, inaddition to a conducive political environment that will permit industrialization to take place, there is every chance that these emergent groups of companies would do better. Unfortunately, there was nothing to suggest, at the time this study was conducted, that the state really had plans to support the effort of these companies. What is the structure of the downstream sector of the Korean steel industry? Is it different from Nigeria's?

4.2.3 South Korea

With particular respect to the iron and steel sector, the Japan Steel Company that operated in the Korean peninsula during the period of Japanese colonialism, was only engaged in aspects of the upstream operations like making crude steel and casting billets which were later exported to the factories in mainland Japan. As explained earlier, there was no major industrial activity in the part of the peninsula that came to be known as South Korea. However, there were evidences from archaelogoical sources from South Korea that family-owned cottage forge shops that existed in the peninsula, used metal scraps to fabricate crude farming implements like cutlasses and hoe/plough, and cooking utensils before Japan imposed foreign rule on the peninsula. With the advent of Japanese colonialism in 1910, however, they were all wiped out by the policy and actions of the colonial state as part of its agenda to protect the local market for the imported intermediate and capital goods from metroplitan Japan. To a large extent, this was similar to the colonial economy of Nigeria under British foreign rule. What is important to note is that, before the outbreak of the North-South Korean war in 1950, there was really no steel company of significance in South Korea much more talking of its upstream sector.

Rather, what existed particularly after the end of the Korean war in 1953, was the emergence of a handful of Korean merchant capitals some of which had engaged in the sale of imported Japanese goods in Korea during the colonial period. They stocked and sold steel products like rods, simple bars, and coils, but lacked the initial capital outlay required to start off their private steel rolling mills. The same was also true of the local artisans who wanted to set up small foundries where some of the parts of the production machineries of the few textile companies owned by Koreans and operating in the country could be cast. In other words, virtually all the intermediate and capital goods that were consumed in Korea between 1945 and 1960 came in as imports, and were largely handled by the subsidiaries of Japanese transnationals notably, Mitsui and Mitshubisi. Again, this was largely similar to the experience of Nigeria during the two decades after its politcal independence in 1960 in the sense that, only a tiny percentage of hitherto imported intermediate and capital goods was reproduced locally by the subsidiaries of the metropolitan transnationals, while a lion share of these goods were still imported. Structurally, therefore, there was a higher import-dependency ratio in what one could call a fragile downstream sector of the Korean steel industry than its counterpart in Nigeria. The pertinent question is: How was Korea able to redress the imbalance in its steel industry, while Nigeria lagged behind?

As pointed out earlier, under then General Park military government, the Korean state underwent a significant reconstitution both in terms of its managers, production systems, economic policy particularly industrialisation, among others. For the first time, then General Park's administration, made the majority of Korean publics and all those interested in overturning the table of the country's economic underdevelopment, that it was the collective responsibility of Koreans and nobody's. That notwithstanding, the Korean government was equally very concerned with how to deal decisively with the root cause[s] of Korea's deepening economic backwardness. Without much delay, the Park administration formulated economic policy that was centered on growth. Because of the general backwardness of virtually all sectors of the Korean economy as of 1961, it was natural to expect that the state would be at the heart of all economic policies and programmes all of which were aimed at bringing Korea's beleaguered economy back to a sound footing.

As it pertains to Korea's industrial underdevelopment, the Park regime was of the view that, in order to overcome it, the country needed a big integrated steel company which should be owned by the state. The absence of an enterprising local business class made all the important for the state to intervene. This partly explains why POSCO, the largest and only integrated steel company in Korea, was state-owned. It was after POSCO had come on stream that the downstream sector of the Korean steel industry really began to develop. That is not to assume, however, that some intermediate and capital goods-producing companies did not pre-date POSCO. Let us begin the discussion with these companies.

Before POSCO came on stream, the Park administration had, in a bid to reduce Korea's high import dependency ratio for light consumer, intermediate and capital goods, devoted the first and second five-year economic development plan periods spanning ten years, 1962-1972, to encourage selected local private companies to produce most of the hitherto imported goods. To ensure a success of the import substitution strategy in the Korean economy, his administration marshalled out liberal economic policies that permitted indigenous Korean companies to import production machineries into the country, while imposing policies that made it difficult for other countries to export these goods to Korea. The first group of the target light manufacturing industries consisted of Daewoo, Hyundae, Lucky Goldstar, among others, and some of which became the big business groups of today in Korea. They were involved in the production of basic consumer goods like toiletries, textile, and shoes. As a policy, the Korean government wanted to have its people first clothed adequately, particularly if economic development would be achieved. In other words, the success of the textile industry would go a long way to herald the country's economic development. The majority of Koreans took it as an act of nationalism by buying Korean-made clothes and shoes, despite their poor quality in comparison with those made in Europe. In fact, for the majority of Koreans to voluntarily buy Korean-made goods inspite of their sub-standard quality, helped secure the local market for the Korean textile companies. By 1965, Korean companies like Daewoo and Hyundae were able to produce enough textile and shoes for the local needs of the Korean people, and also started exporting to some of the countries in East Asia like Vietnam.

Between 1965 and 1970, there was a dramatic increase in the demand for both intermediate and capital goods in Korea. For instance, *per capita* steel consumption rose from 20kg to about 175kg with steel beams, flat sheets, rods, coils and bars dominating the imported steel products. The increased demand for these products stemmed largely from the rise in real estate development, and the construction of roads and bridges among other infrastructural facilities. In fact, the massive construction of infrastructural amenities was part of the government's broad policy of opening up a hitherto rural country for economic development. The importation of completely knocked down parts particularly those used for the assembly of production machineries equally witnessed a rise in demand. The same was true of the establishment of light metal fabricating industries, and steel rolling companies like the Dongguk Steel Company, which rolled out steel rods locally, using imported steel billets most which came from Japan.

As noted, by the end of the second economic development plan period, it was quite obvious to the Korean government that, though much effort was made to reduce Korea's import for light consumer goods, little or no progress was made to curtail the importation of intermediate and capital goods, particularly as there were no companies in Korea that could produce them. Like the Latin-American countries' disappointing experience with the import substitution approach in the 1950s, Korea's intermediate and capital goods-producing companies were still deeply import-dependent on Japanese companies for their basic raw materials, production machineries and their spares. In other words, until the limitations of the import substitution approach was resolved, the Korean economy would continue to be disarticulated in its quest for industrialization. It was partly in an attempt to transcend the limits of the import substitution strategy that compelled the Korean state to establish POSCO as noted earlier. If the companies that came to constitute the downstream sector of the Korean steel industry in the period before the establishment of POSCO were wholly import-dependent for their production machineries, spares and basic raw materials, the question that arises therefore, is whether there was any change in the structure of the downstream companies since POSCO came on stream. Obviously, there was a significant change in the structure of the companies operating the downstream steel sector of Korea.

The main policy thrust of the Third Five-Year Economic Development Plan of the Korean government as explained earlier, was to 'deepen Korea's industrialisation'. That is, to overcome the limitations of the import substitution strategy by having companies that would not only be relatively self-reliant in terms of their basic raw materials, but also, have the capacity to really control Korea's industrialization. How did POSCO help to achieve this policy objective?

Five groups of companies dominated Korea's downstream steel sector after POSCO was established in 1968, namely, automotive, machinery, ship-building, fabrication and others. Each of them had specific target market in the local economy. As companies engaged in the downstream operations, the Korean government had planned that they would rely on POSCO for virtually all their basic raw materials needs.

In order to reduce, or if possible, totally eliminate the high import dependency of the intermediate and capital goods-producing companies in Korea, POSCO had a wide range of product profile. Furthermore, the government had adopted a policy of increamental replacement locally, of imported production machineries, their spares, and raw materials all in a bid to have a sustainable industrial growth in Korea. Beginning in 1973, the government's policy of increamental substitution of steel and steel related hitherto imported intermediate and capital goods, was to be fully achieved by 2010. In essence, over a certain period of time, it was expected that a particular kind of steel product should no longer be imported, but produced locally. So far, South Korea was able to use this policy to attain a reasonbale level of self-sufficiency in some of the steel products consumed in the country. This is quite instructive for Nigeria. Details on this later.

As it stands, POSCO was perhaps, Korea's real hope of launching the country's technological age. Not surprisingly, the activities of the company straddled both the up- and downstream operations of Korea's steel industry. In the downstream sector for instance, POSCO's ancilliary companies like POSMEC produce heavy machineries and as well, engaged in high-tech industrial engineering. Others like POSEC, carried out major engineering and construction works, while POS-M manufactured and maintained parts of the blast furnace. In fact, POSCO was, at the time this study was conducted, involved in all the production of autoparts, heavy machineries, shipbuilding and fabrication of major parts of heavy equipment. How well the company had performed will be treated in the next chapter.

The ultimate aim of the big business groups, as explained earlier, was to accomplish the state's policy of export-oriented industrialisation. They were also engaged in virtually all sectors of the economy. In the steel sector, the *chaebols* produced basic raw materials some which were capital goods themselves, or serve as inputs for manufacturing them. Hyundae Shipbuilding Company, for instance, relied on POSCO for special flat that it uses for constructing ships. The same was true of both Hyundae Motors and Daewoo Motors, which sourced flat steel products from POSCO for the purpose of building car/truck bodies. POSCO's Foundries also supplied most of the engine blocks for these auto-industries, with the rest coming in as imports. Most of the steel rods, beams and bars used by Hyundae Construction Company within and outside South Korea, were largely supplied by POSCO.

Unlike in Nigeria, where there was hardly any linkage between the steel companies and heavy industries, the situation in Korea was quite different. In Korea, for instance, not only did POSCO establish links with the heavy and chemical industries in particular and the rest of the economy generally, the company was poised for the reversal of the deepening import dependence of the capital goods sector of Korea. Whether it was able to achieve this goal will become clearer when its performance is discussed in the next chapter.

One major development in the downstream sector of Korea's steel industry since the advent of POSCO, was the establishment of many small steel and steel -related companies that were largely dependent on it for their basic production inputs. With little internal self capacity to meet some of their raw material need as well, these companies as shown in Table 3.4, produced most of the spares for the automobile industry, and for other companies that were engaged in manufacturing heavy machineries, ship-building and construction. As can be gleaned from the above table, the majority of the companies operating in the downstream steel sector, were set up in line with the state's policy and programmes on industrialisation. That is, to help satisfy locally, the basic raw material needs of companies involved like civil construction and the manufacturing of heavy machineries. Largely dependent on POSCO for their basic raw materials, some of these downstream companies like Kangwon Industries Limited, Chongno-Ku, Seoul; Hankook Steel & Mill Company Limited, Kyongsangnam-do; and Dae Han Steel Mill Company Limited, Pusan, produced steel rods and bars, angles, beams and channels, stainless steel pipes, rails and track shoes for the different companies engaged in these activities.

All this was in sharp constrast with the situation in Nigeria where the few companies operating in its downstream steel sector, were not only 95 import-dependent for their raw materials, but also, dominated by private foreign capitals. Furthermore, the Korean state was of the belief that import substitution strategy could be realised instalmentally. Hence, Korean companies operating in the downstream sector of the country's downstream steel sector was planned to help reduce its import dependency ratio for intermediate and capital goods over a long period of time. As pointed out earlier, the Korean government had, in 1973, planned that twenty years later, that is, by 1993, South Korea would be wholly self-sufficient in the production of basic constructional steel products like rods, bars, angles, beams and channels. Subsequent regimes after the overthrow and death of General Park in 1979, continued with the national industrial/steel policy. Based on the official sources from the Korean government and the interviews the author had with officials of POSCO and the Korean Development Bank in Seoul in June 1997, Korea attained 100 percent self-sufficiency in basic constructional steel products like bars, coils, beams and channels in 1995.[15] Korea's drive towards self-sufficeincy in steel products was the exact opposite in Nigeria, where the state was itself, a clog in the wheel of the progess of the country's steel sector. As of 1998, 98 percent of simple constructional steel products like rods, coils and bars used in Nigeria were all imported. Details on all this will be treated in the next chapter.

4.3 Concluding Remarks

Both Nigeria and South Korea inherited an underdeveloped economy inclusive of its steel sector from British and Japanese imperial rule respectively. This was so because the colonial state powers never had on its agenda, the industrialisation of the two former colonies. In the post colonial period, however, they had contrasting experiences with industrialization. In the area of steel development, while Nigeria is yet to get it really underway, South Korea recorded a relatively impressive growth. In all , they were not without problems. It is the performance of the steel industry of both countries that will now address in the next chapter.

Notes and References

1. Based on the author's field trips to the Steel Department of the Federal Ministry of Power and Steel, and the state-owned steel companies at various times in 1996.
2. *ibid.*
3. *ibid.*
4. *ibid*
5. Based on the author's checks during his trips to SRMEA in 1990 and 1995.
6. Based on the author's field trips to the steel companies.
7. Based on the author's researches at the MAN offices in Lagos.
8. Author's checks at POSCO's Steel Works at Pohang and Kwanyang in 1997.
9. Contrary to the view that countries lacking in iron ore and coal should have no business in steel development, the Korean government was of different opinion. Its officials were of the view that, its strategic importance makes it compulsory for developing countries that really wanted to reverse their industrial backwardness to have a viable steel industry. In essence, the success of steel project in developing countries is tied largely to the commitment of the state to steel in particular, and industrialisation generally.

10. Based on the author's checks at POSCO.
11. Based on the author's interviews with officials of KOSA.
12. Based on the author's field trips to the company in 1988, 1990 and 1996. The unseriousness of the Nigerian government with steel development unfortunately turned MTL into a small foundry shop when it began operation.
13. Based on the author's numerous interviews with Bunu Sheriff Musa who was Minister for Mines, Powere and Steel 1986 and 1989. During this period, the author was attached to the Minister as a special correspondent, and in that capacity, had access to vital information on the steel industry.
14. Based on the author's checks from the steel companies in 1990, 1993, and 1996.
15. In June 1997, the author had a Roundtable with the officials of POSRI in Seoul. The discussion centered on crucial industrial policy issues particularly on steel. The officials of POSRI made vital statistics on the Korean steel industry available to the author. At other times, too, the author had series of arranged private and official talks with officials of the Korean government and officials of KOSA on the country's steel industry in particular and the economy at large.

Chapter Five

Performance of the Steel Industry

5.1 Introduction

This chapter attempts a critical and comparative assessment of the performance of the public and private steel companies in both Nigeria and South Korea, paying particular attention to the their stages of growth, capacity utilisation, product mix and the linkage of these companies with the intermediate and capital goods-manufacturing companies in particular, and other sectors of the economy generally. Before evaluating the level of development of the steel industries of Nigeria and South Korea, it is in order to explain some of the basic unsettled terminologies that will enhance our understanding of the comparative analysis of the performance of the steel sectors of both countries.

5.2 Explaining Basic Unsettled Terminologies Underpinning Comparative Analysis of the Performance of the Steel Industries in Nigeria and South Korea

Like other sub-sectors of the economy of a country, the steel industry has its own life cycle that is broken into various stages of growth. Which is why, if a country's steel sector is said to have attained market maturity, it means that it has passed through the different stages of growth, with each stage giving an important insight into the level of the country's industrialisation generally.

Six stages of growth can, at least, be discerned in the life cycle of a country's steel industry, developing or developed. They are: i.] the **take-off stage;** ii.] **early growth;** iii.] **late growth;** iv.] **market maturity;** v.] **declining period;** and vi.] **stability.** This is graphically represented in Figure 5.1, located in the Statistical Appendix. Since each stage of growth exhibits features that best expressed technically, it is in order to first clarify these terminologies. They are: self-sufficiency ratio in steel; product mix; *per capita* steel consumption; and total local sub-sectoral steel consumption.

The **Self-Sufficiency Ratio** [SSR] of the steel sector of a country is an expression of its total internal capacity to meet its overall domestic crude steel demand. Calculated in percentage points, it is as an expression of the total domestic production over the total steel demand. That is, **SSR = domestic production / total steel demand expressed in percentage,** with 100 percent taken as the cut-off mark for self sufficiency. In other words, a country with an SSR of 100 and above is considered to have a well developed steel industry.[1] For instance, as of 1996, Japan had an SSR of 121 percent compared to Korea's 99 and Nigeria's 25. See Table 5.1, located in the Statistical Appendix for details. No doubt, these figures display the different levels of development in the steel industries of these countries. However, it is important to caution that the SSR index alone, cannot be used to determine the overall level of development of a country's steel industry. The amount of steel consumed per head is equally important.

Per Capita **Steel Consumption** [PCSC], is an expression of a country's total quantity of steel consumed over the total population in kilogrammes. The International Iron and Steel Institute, Brussels, [IISI], Belgium, has put the world's minimum average of PCSC at 133 kg, which means that countries with a PCSC above 133kg have a relatively developed steel industry while those with figures below it are considered to have an underdeveloped or backward steel sector.[2] For most of the industrial market economies like Britain, Germany, France, USA and Japan, the PCSC ranges between 600 and 800 kg, while the bulk of the developing economies like Nigeria has a PCSC of 30kg. Important as a country's PCSC might be, it does not alone, give the real status of its steel industry. For instance, though Korea had a PCSC of 600kg like most of the industrial economies as of 1996, it does not mean that its steel industry had attained market maturity; nor should it be assumed to be on the same level of development with those of major steel producers like Germany and Japan. For, Korea's steel sector was still at late growth as of 1998. The rise and fall of the PCSC of a country over a period of time, is also partly indicative of the overall performance of its economy. See Table 5.2, located in the Statistical Appendix for details.

The **Product Mix** [PM] of a country's steel industry gives a range of the steel products that are manufactured locally by its steel companies. As an index for measuring a country's level of industrialisation, the PM shows the level of local content in the production of light, intermediate and capital goods. That is, the extent to which the products of the steel sector are used in manufacturing these goods.[3] Which means that, if a country's steel companies' PM consists mostly of simple structural steel products like rods, coils and bars as in the case of Nigeria, there is every tendency for its economy to be largely involved in preliminary construction of some basic social infrastructures like roads, bridges and public buildings, and little or nothing in the process of manufacturing. Suffice it to say that, the PM of country's steel companies gives some useful insights into the structure of its manufacturing industry.

The **Total Sub-Sectoral Steel Consumption** [TSSSC] expresses the quantity of steel used by the various industrial subsectors of an economy as a percentage of the total steel produced locally over a period of one year. As one of the major yardsticks for determining a country's level of steel and industrial development, the TSSSC measures in percentage point, the share of the local steel produced and consumed by both the construction and manufacturing companies. Putting the cut-off mark at 50 percent, the IISI stipulates that a country whose construction industry's share of the TSSSC is above 50 is an indication that, not only is it relatively a growing steel sector, but also, a developing economy. On the other hand, countries with a TSSSC of 50 and below in the construction industry suggests a relatively more mature steel sector and economy generally.[4] Which means that Nigeria's construction industry TSSSC of 90 percent is an indication that, its steel industry in particular and economy generally, are far less developed than Korea's with a TSSSC of 54 percent. The situation is different in the manufacturing industry where a higher percentage of TSSSC is indicative of a more developed economy. The IISI puts the minimum TSSSC of a country's manufacturing at 50 percent, meaning that higher it is, the more industrialised its economy would be. Having clarified these concepts, let us explain the various stages of growth of the steel industry.

5.2.1 Stages of Growth of the Iron and Steel Industry

i] Take-off Stage: At this stage, a country's steel industry, particularly its upstream sector, has companies that manufacture steel products, and as well; produce iron ore, coal and alloying minerals. Though a high premium is usually placed on the production of simple constructional steel products given the relatively overall backwardness of the economy at that level, there is really no hard and fast rule that a country must bias its product mix for simple constructional steel products. On account of British colonialism, steel development first began in Nigeria with its downstream sector instead of starting with its upstream sector. For, if it had begun with its upstream sector and the right leadership, it would have helped influence the structure of the companies engaged in the downstream operations. However, the past experiences of developing countries like South Korea showed that the trend could be reversed. That is, redressing the imbalance in the country's steel industry. The SSR of countries at this stage of growth, is usually within the range of 1-30 percent, the bulk of which comes from simple constructional steel products. The construction industry's percentage share of the TSSSC is usually in the range of 98 percent with the remaining two perecent left for the manufacturing sector, which depends largely on imported spares for the local repair of broken down machineries. It is possible for a country to take-off wrongly with its steel development, in which case, there is really no steel policy guiding its steel projects. In such situation, large steel companies could be established. So, too, would other companies in the downstream sector. But one of the defining features of such steel industry is its incoherence. That is, the steel industry in both up- and downstream operations, will virtually be import-dependent for basic raw materials and spares even when they abound locally. This is clearly the case for Nigeria and Zimbabwe, two major steel producing countries in Sub-Saharan Africa. Next, is the early growth stage.

ii.] Early Growth: If a country's steel sector takes off with the right macro and micro-economic fundamentals and a cogenial political environment that will permit industrialisation to take place, its steel industry is naturally expected to witness some early growth. At this stage, both the nature, quantity and quality of the steel products rolled out of the mills, and their consumption locally, should be expected to increase significantly. Like at the previous stage, where over 80 percent of the total finished steel products is usually used by the construction industry, at the early growth, it should be about 70 percent. At this stage too, it is expected that the PM of the steel companies would have a wider range since the local economy should be expected to expand particularly in terms of industrial activities. Because of the rise in real estate development and more construction of social infrastructure expected in the economy, there would be an increase in the domestic demand for steel products at this stage than the previous one. All this would, in turn, cause a dramatic increase in the tonnage of the total crude steel produced annually over the figure at the take-off stage. Not surprisingly, the SSR ranges between 40 and 50 percent as against 30 at take-off; and the percentage of the construction industry of the TSSSC hovers around 85. This is the stage which China's steel industry was as of 1998, which is quite far ahead of Nigeria's.

Compared with Nigeria, China's steel industry is relatively more developed. But that is not to assume that the steel sub-sector of China's economy took off rather smoothly. Rather, it is being beset with problems right from the onset due largely to its socialist ideology, which kept China's steel sector out of the steel industry of the global market economies of which Nigeria is part. Even with the recent attempt by China to embrace the market forces while still retaining its socialist ideology since the turn of 1990s, its steel industry and other sectors of the economy are still centrally controlled. This tends to obscure the significance of using China's steady reduction in the amount of steel consumed in the construction industry as a rigorous index to compare it with Nigeria where it is rising. Furthermore, there was remarkable diversification in the PM of China's steel companies coupled with the rise in its SSR from 30 to about 68 in the last one decade, and an increase of its annual crude production to 2 million metric tonnes from less than a million in the recent past. On the surface, all these indices would naturally have pushed China's steel industry to the level of late growth where the Korean steel sector was as of 1998. In reality, China's steel industry is still witnessing early growth and is far behind Korea's. For, what actually took place in China in the last ten years especially since the end of the cold war in 1989, was largely a sub-sectoral jolt in the country's industrial sector, spreading across the larger economy especially as it pretends to abandon its communist ideology and embrace market forces

iii.] Late Growth: At this stage, the steel sector of a country would have started showing some signs of coming of age both at the local and global levels. It does this internally by impacting considerably on the pattern and structure of manufacturing of the country's economy, particularly in helping to produce and manufacture locally, most of the basic inputs for the production of machines and their spares, and other intermediate and capital goods. It also shows a noticeable shift in the srticture of manufacturing as emphasis would be placed on real manufacturing and the assembly of intermediate and capital goods in the country. So, too, is it expected that the gains of added value in the industrial sector would begin to accrue to the local economy. Indicative of this stage also, is the effort to put the overall economy on the path of sustainability. Some of these signs have manifested in both the steel sector and the larger economy of South Korea.

Evidently, Korea's steel industry is at the stage of early growth. With its first product rolled out in July 1973, the government-owned POSCO has, since then, grown by leaps and bounds. Together, POSCO and other Korea's private steel companies were able to increase the country's total output of crude steel from about 100,000 metric tonnes in the early 1970s to 36 million metric tonnes in 1996. From a narrow range of simple constructional steel products in the early 1970s too, the PM of Korea's steel companies was already in hundreds by 1996. Infact, some of the steel products rolled out of the steel companies are themselves, either intermediate and capital goods, or serve as basic inputs for the production of these goods. This is clearly shown in Table 3.4. That is not all, as Korea's SSR of 20 percent in the 1970s had risen to

99, next to Japan's among the Asian NICs. Comparing all this with other Sub-Sahara African steel-producing countries like Nigeria and Zimbabwe, Korea is quite far ahead. See Table 5.1 for details in the statistical appendix. However, while it is not in doubt that Korea's steel companies have helped structure the intermediate and capital goods-producing companies, the percentage shares of the country's construction and manufacturing companies in its TSSSC were still 58 and 42 respectively. What all this means, is that, Korea's level of industrialisation is still youthful.

All the same, Korea's steel industry had made tremendous impact on the growth and of the global steel industry by contributing 5.32 percent to the its total output of steel crude in 1996. From a net importer of finished steel products in the early 1960s, Korea also rose to become the greatest exporter of steel products to countries like Vietnam among other South East Asian countries. Within the period of two decades too, POSCO had signed joint-venture agreements with other steel companies in the rest of the world, particularly in the United States of America and Europe. In other steel producing countries like Australia and Canada, POSCO had also jointly invested in the development of iron ore and coal. So far, one has followed convention. A critical analysis of all this will be undertaken shortly.

iv.] Maturity: At market maturity, a country's steel industry ought to have overcome virtually all the problems identified at the stages of take-off through early growth to late growth. It is particularly expected to be so as the country ought to have fully industrialised as well. The implication of all this for the SSR is that, it will not be less than 100 percent; and the manufacturing industry's share of the TSSSC will be about 90 percent with the remainder going to the construction industry. In fact, the percentage share of the construction industry in the TSSSC is expected to decline continuously since most of the social infrastructure and larger part of real estate development, would have all been completed at this stage. In order to remain in business, the steel companies are expected, at this stage, to begin to diversify into other businesses that are either directly related to the steel industry like the sale, installation and maintenance of heavy equipment; or others that are remotely connected to it like information/data science. Most of the traditional steel producing European countries like Britain, France, Austria, Germany and

Belgium had attained this status in the 1960s. In fact, VOEST ALPINE of Austria had long specialised in the production of the concast machines after the Austrian steel industry attained maturity. Japan's steel industry entered into market maturity in the late-1970s, and has since then, had its steel transnationals diversified into the non-steel sector.

v.] Decline: As to be expected of an equilibrium point, the period of market maturity is usually difficult to sustain over a long period of time without sliding into decline. Why? It partly stems from the inability of the local economy of the steel producing country to absorb most of the steel products and vagaries of the global steel market. This is all the more so because at maturity, it is expected that the overall steel demand in the economy would begin to decline following the completion of most of the construction of social infrastructure. The only other major consumer of steel products left, would be the manufacturing industry. But even then, its percentage share of the total steel consumed locally is usually not enough for the steel companies to maintain their previous momentum of activity in order to post profits. Inevitably, decline will set in. And this was the situation in the steel industry of the market economies in the early 1970s, particularly after all the repair and rehabilitation of war-torn European countries damaged during the World War 11 were completed. Not only that, as some of the major sectors of the global economy that consumed a lot of steel products such as ship-building had started to face market decline since the turn of the 1970s. As for the United States of America for instance, there was really no need to continue to invest in the steel industry especially as it had met its domestic steel demand.

Faced with declining steel market, the majority of the European steel producing countries had no option than to make frantic efforts to maximize the highest returns on investments as the market outlets for steel products, and investment capitals within Europe had become overconcentrated. In other words, the survival of the European steel companies in particular, and the prospects of the global steel industry generally, lay outside of Europe. Rather, it is in the developing economies whose steel industry is either at take-off stage as in Nigeria, or at late growth as in the case of South Korea. Perhaps, this partly explains why the European steel companies have cornered the sale of production machineries and their spares as well as servicing them in the emergent steel producing countries in order to remain in business. The author has dwelt extensively on this eslewhere and should not detain us here.[6]

Some of the major features of the steel industry of a country under decline, are the closure of major production lines, placing ceiling on the entry into, and exit from the industry; imperfect market competition; imposition of production quota; privatisations, mergers and acquisitions; and bankruptcies and liquidations. Where the state intervened into the process of steel development without a clear-cut policy, these problems could also arise in its steel industry, though it is still at the stage of take-off as in the case of Nigeria. Care should be taken not to construe this as problems emanating from decline as a result of the inability to sustain market maturity. Other basic indicators like the SSR, PM and TSSSC do not really change much since it had already attained market maturity. Of the European steel producing countries that went into decline in the 1960s, Germany appears to be one of the few that has survived, while Britain, apparently unable to manage the British Steel Company, sold it to private capital. As for France, its steel industry is still far from being well, but barely alive. Drawing on the past experiences of the European steel producers at market maturity, Japan diversified its steel industry as soon as it began to notice some signs of decline. Perhaps, that partly explains why its steel industry is still kept afloat inspite of the glut in the world steel market since the mid-1970s.

vi.] Stability: At issue, is not how to get to this stage, but maintaining it. It is about the most difficult stage to maintain in the steel industry because of the unpredictability of the market forces and the contradictions stemming from monopoly capitalism all of which help to upturn its equilibrium. One of the major features of countries whose steel industry have attained this status, is the ability of the steel companies to fully have diversified into other businesses that are unrelated to steel ostensibly to complement their traditional operations.

Some of these businesses included stock trading, acquisition and regeneration of ailing companies, manufacturing of information/data equipment, space science. Germany can arguably be said to be the only country whose steel industry has attained this stage, having survived the periods of both recovery and decline in the 1960s. One thing to note about this stage however, is the difficulty of remaining there for long since a jolt in both the local and global steel markets is likely to upstage its stability. With this analysis in mind, let us now turn to the evaluation of the performance ot the steel companies in Nigeria and South Korea.

5.3 The Performance of Both Public and Private Steel Companies in the Context of Capacity Utilisation, Production of Intermediate and Capital Goods, and Linkages with other Sectors of the Economy

In evaluating the performance of both public and private steel companies in Nigeria and South Korea, this section pays particular attention to their capacity utilisation and role in the production of intermediate and capital goods, and overall linkage with other sectors of the economy. Two reasons can be advanced for this. First, the capacity utilisation of the steel companies gives an insight into the level of industrial activity in that sector. Second, most of the products of the steel companies are not only intermediate and capital goods themselves, but also, serve as basic inputs for manufacturing machines and other heavy equipment including their spares. In essence, they are indicators of a country's level of industrial development. Let us begin the discussion with Nigeria's experience.

5.3.1 Nigeria

As noted earlier, Nigeria's steel industry was, at the time this study was conducted, still dominated by the state-owned steel companies with a small fraction controlled by a combination of local and foreign private capitals. The final products of the steel companies, public and private, were essentially simple constructional steel products like rods, bars and coils. They ranged between 0.05mm to 50mm in diameter and with a maximum length of 12 meters. Though the DSC was initially planned to be fitted with production lines that would roll out more sophisticated constructional steel products such as tees and channels, angles, I-beams, it was only the facility for producing tees that was installed but broke down during trial production in 1982. It was not repaired and eventually scrapped.

As an integrated steel plant, DSC was built with facilities to process iron ore first into steel, then cast into billets, and finally into rolled products. Its pellet plant had the capacity to produce 1.5 million metric tonnes of oxide pellets from iron ore annually. However, since inception in 1982, about 20 percent of its installed capacity had been achieved due essentially to the inability of the company to source for funds with which to import iron ore and other consumables. The direct reduction [DR] plant on which the whole steel technology at the company was based, had a yearly capacity to produce 1 million metric tonnes of crude steel. Like in the pellet plant, the DR plant hardly recorded more than 20 percent of its capacity right from its inception till 1996 when the company was shut down indefinitely. Even at that capacity, this percentage was no less than an average of six skeletal productions in a year, meaning it was not really representative of the activity in the company.[7]

Equipped with four electric furnaces, DSC had capacity to produce 1million metric tonnes of crude steel annually. Because these facilities were built with refurbished parts, they already had a shortened life span from the onset. For instance, they had five as opposed to 15 years of life span. Not surprisingly, two of the furnaces fell into disuse in 1990 inspite of minimal production, meaning that the plant's capacity was already redued by 50 percent. Of the remaining two, one of them had a major fault in 1992 and was yet to be repaired before the company was shut in 1996. In essence, it was only one furnace with a capacity to produce 250,000 metric tonnes of liquid steel that was functioning. It only produced 20 percent of this capacity before it finally broke down in 1995. With the four furnaces broken down, the government had to close the company in 1996.[8]

Before the company was closed down indefinitely in 1996, DSC had an installed annual capacity to cast 960,000 metric tonnes of billets, a semi-finished steel product. This facility was built to cast billets continuously throughout the year, but it produced only at 20 percent of the installed capacity. The DSC also had an in-house steel rolling mill with an annual capacity of 320,000 metric tonnes, and whose billet need was expected to be met wholly from within the company. However, only 19 percent of the rolling capacity of the in-house mill was met. 10 percent of its rolling mill's billet requirement could only be achieved internally. Even the 19 percent performance was limited to a small range of products, mostly steel bars and rods which were fairly in high demand locally. The lime plant had capacity to produce 66,000 metric tonnes of burnt lime yearly. For one thing, lime is a major additive used for making steel, and the DSC started operation procuring the product from the Cross River Limestone Company [CRLC] located at Mfamosing. The plant produced at 80 percent of its capacity with about half of it used by DSC, while the remainder was sold to other local consumers. This capacity was achieved before the Nigerian state embraced the World Bank/IMF-led adjustment programme in June 1986. Partly in complaince with the Bank/Fund's conditionalities for loan, the Babangida's administration sold out the DSC's share in the CRLC to local private capital. Since the advent of adjustment, the CRLC could only produce skeletally. With the CRLC yet to recover as of 1998, what it all means is that, with the recovery of the DSC, if it ever does, the company might as well begin to source for limestone at its own cost.

DSC's foundry was planned to be captive to its various production plants in the sense that it was expected to produce locally, some of their basic spares. It did not really have an installed capacity. Rather, it had facilities for fabricating some intermediate goods for the plants as well as for the support service sectors of the company like the real estate. Among the notable intermediate and capital goods that were fabricated at the DSC's foundry were crawsher balls, hammers, buoy sinkers, pump impellers, brake drums and shoes, connecting rods for machines, valves, cam and crank shafts. Although the company's foundry fabricated these goods, most of them were not for internal use, but were produced on customers' specification and request. Some of the companies that patronised DSC's foundry included: the Nigerian Railway Corporation, and the Nigerian Ports Authority. Why did the DSC's foundry decide to produce for commercial purposes? With the state subvention to the public steel companies among others, first drastically reduced and then began to dwindle since the state adjusted in 1986 until it was suspended indefinitely in 1996, the company had no option but to use its foundry to raise some fund. The implication of this, is that, not only has it amounted to the cannibalisation of the entire steel project, but in particular, resulted in the early break down of the foundry arising from its excessive use. Not unsurprisingly, the DSC's foundry fell into disuse in 1994 before the company was shut indefinitely in 1996. If the situation at the DSC was that bad, it was worse at the steel rolling mills that were planned to wholly depend on the former for their billets needs.

Each of the public inland steel rolling mills at Katsina, Jos and Oshogbo had an annual installed capacity to produce 210,000 metric tonnes of rolled long and round products. The product mix of the steel rolling companies was the same: 0.05-50mm in diameter, with length not exceeding 12 meters. It was only the Katsina steel rolling company, however, that had facility to produce products of diameter below 0.05mm. That is annealed coils which are meant specifically for use by the wire drawing companies. This production line was only in skeletal production for about two years before it was finally shut down in 1989 due largely to lack of patronage locally. On account of special arrangement between Katsina Steel Company and Kobe Steel of Japan, the company, unlike its counterpart at Jos and Oshogbo, was able to enjoy a fairly constant supply of billets. As a consequence, the company was able to achieve 40 percent capacity utilisation on two major products, 0.25mm long, and 0.25mm round; but not across the entire product profile. Forty percent capacity utilisation was the company's peak performance before the advent of the structural adjustment programme of the World Bank/IMF in 1986; thereafter, skeletal production set in. Even uptil 1991 when the Technical Committee on Privatisation and Commercialisation [TCPC] and later known as the Bureau of Public Enterprise [BPE] proposed to privatise the company, there was no improvement. All this caused the withdrawal of the financial aid by Kobe Steel of Japan to the company. Since 1991, Katsina Steel Rolling Company had hardly exceeded an average of 20 percent of its installed annual capacity. Even with this capacity, production was really no less than 60 days out of the 365 days in a year. With the abandonement of the privatisation project of the steel rolling mills by the state in 1993, Kobe Steel of Japan, again, reactivitated its bilateral finanacial links with Katsina Steel Company. Why the renewal of the link? It is partly because Kobe Steel was the foreign contractor company that built the company. Much more importantly, is the fact that, as a relatively 'new comer' into the Nigerian economy since 1960, the Japanese government has been using major Japanese companies like Kobe Steel not only to promote closer economic ties with Nigeria, but also, to help gain strong access into the large and untapped Nigerian steel market. As to be expected, Kobe Steel gave US$1.6million to the company in 1993 as loan to refurbish its aged furnace. The loan was, however, converted into billets largely at the instance of the company's management and not really that of Kobe Steel. With the billet, the company was able to operate in 1994 at an average capacity of 30 percent of its two products already mentioned above.[9] Following the gross human rights abuse by then Abacha junta, Japan joined some other member-countries in the international community that isolated Nigeria, thereby compelling Kobe Steel to suspend further bilateral links with the Katsina Steel Company. As of 1995, Katsina Steel Company was already severely financially constrained that it could not procure billets; and by the following year, the Nigerian government closed all the public steel rolling companies indefinitely.

The performance of the Jos Steel Rolling Company with respect to its capacity utilisation was never encouraging right from its inception in 1983, until it finally closed shop in 1996. Between 1983 and 1986, its production was an average of a mere 15 percent of its installed capacity, the company's fundamental problem being lack of funds to import steel billets as the DSC, which was to meet all its billet demand could not produce. With the advent of adjustment, the situation worsened due to the sharp increases in the price of imported billets. Even after the official termination of adjustment in 1994, the situation did not improve. For instance, throughout 1990, the company did not produce for more than four times. In 1993, its mill was not in production at all. In 1994, it had its first production in March using billets that were supplied by one of the company's major distributors. The company only charged fees for rolling the products. In fact, this was gradually becoming its source of raising fund since it could no longer import billets. Though the company's foundry was made captive to the mill, it was not fitted with facilities like that of DSC. Like its counterpart at Katsina, therefore, it could not forge any of the major spares of the furnace except simple bolts and nuts. That partly explained why it was also not commercialised.

Of the three state-owned steel rolling companies, the Oshogbo Steel Company was the worst performer. Within the first three years after it was commissioned in 1983, it recorded an average of 10 percent capacity utilisation. Between 1986 and 1990, there was no improvement. In the next two years, that is, 1991 and 1992, the company did not roll out any product at all. In 1993, it ran some skeletal operations for four hours because one of its distributors supplied billets which was rolled for a fee. Thereafter, the plant became redundant before the company, together with public steel companies were closed indefinitely as explained earlier. Locate Table 5.3 in the Statistical Appendix.

There was a generalized inactivity which was tied to the very poor performance by the public steel companies. In 1989, for instance, none of them produced steel rods for reinforcement and wires, and this trend continued into the 1990s until the government closed them indefinitely in 1996. The overall consequences of the poor performance of the state-owned steel companies on the country's industrialisation in particular, and its economy generally, were many and grave. First, the major aim of using the steel companies to produce locally, basic intermediate and capital goods hitherto imported, could not be achieved. In fact, a large proportion of both intermediate and capital goods had to be imported by both the steel companies and private heavy industries to meet local demand. Furthermore, the local capacity to fabricate or refurbish these goods locally still remained tied to the fluctuations of the global steel market. For one thing, the global steel market is dominated by the steel transnational corporations that are aggressively seeking for market outlets in the developing economies, meaning that unless the steel producing countries in the Third World get their fundamentals right, their steel industries hardly stand a good chance of survival. Not surprisingly, the structure of imported intermediate and capital goods in Nigeria as shown in Table 5.4, located in the Statistical Appendix, was still dominated by long and flat steel products, steel tubes and pipes, railway tracks, shoe brakes and coaches among others. The situation was not really different from that of Zimbabwe Iron and Steel Company, Zimbabwe, where its capacity utilisation was reduced to five percent in the late 1980s, before the company was later shut. However, by 1997, the Mugabe government had, after receiving some external financial assistance, started the reactivation of the company's flat steel mill that was broken down.[10]

The private steel companies, as noted, had a small capital base with installed capacities not exceeding 50,000 metric tonnes of liquid steel, and about 30,000 metric tonnes of rolled products. Small as their capacities were, they never operated at more than 50 percent. This was even in the 1970s when the average *per capita* steel consumption was 150kg. Operating at less than 50 percent installed capacity, the steel companies could not meet the local steel demand. In fact, if their capacity was weighted against the total steel demand in the country in the 1970s, they accounted for a mere 5 percent. One of the major reasons for their overall poor performance was due to the absence of a steel policy. Without such a policy, the private steel companies were really no less than steel merchants. All this led to their easy collaspe following the advent of the World Bank/IMF-sponsored adjustment programme, particularly as cash squeeze worsened in the economy. Even after the official termination of the adjustment programme by then Abacha administration in 1994, more private steel companies still closed shops with nothing that suggests their recovery in the near future. Worst still, the private steel companies, like their counterparts in the public steel sector, produced mainly rods and wires ostensibly for construction purposes, meaning that they hardly had any useful input into the local production of intermediate and capital goods. This partly explains why these goods were still, and perhaps, will continue to be imported into the country for a long time to come. Table 5.4 clearly demonstrates that most of the intermediate goods that would have been produced locally, if all was well with the steel companies, namely, channels and angles are still being imported. The same was also true of capital goods like flat steel sheets and CKDs. With respect to the value of the import for intermediate goods particularly since 1986, it was a steady increase: 18.3 million naira in 1986; 65.2million naira in 1987; 188.6 million Naira in 1988; 435.9 million naira in 1989; and by 1990, it rose further to 848.9 million naira. As of 1997, it was estimated at 3.5 billion naira.[11] Also, there was a steady increase in the import levels for steel plates and completely knocked down parts for automobiles and other machineries within the same period, 1986-90. See Table 5.4 for details.

From all indications, the combined performance of both the public and private steel companies still shows that not much was achieved in terms of their capacities to meet the country's internal demand for steel products, and intermediate and capital goods. Even at 100kg *per capita* steel consumption in 1984, the steel companies total output still could not meet the demand. In essence, there was a gross undercapacity utilistion across the steel companies, public and private. As noted earlier, though the advent of the Bank/Fund-led adjustment in 1986 had worsened the situation, the real problem with the country's steel sector is that, steel development was not really on the agenda of those in charge of the state. In other words, adjustment cannot be blamed for all the woes of the steel industry. Rather, the impact of adjustment on the steel sector can only be understood under the adjustment situation. With the de-valuation of the naira, production inputs for the steel companies shot up dramatically since they were largely import-dependent for their raw materials, notably iron ore and steel billets. For instance, the landing cost of a ship load of iron ore at DSC as of 1985, was estimated at 1.5 million naira with the capacity of the ship put at 30 tonnes dead weight. By June 1986 when the Nigerian state adjusted, it rose to 4 million naira, almost trippling the pre-adjustment price. The trend of rapid increase in the price of production inputs for all manufacturing industries inclusive of those in the steel sector continued into the 1990s, forcing most of them to close shop temporarily pending the cessation of the crisis in the economy. In fact, as from 1990, DSC had been particularly badly hit by lack of iron ore that it scarcely operated its DR plant. In 1992, it could no longer source the fund to import iron ore due in part to the dwindling subvention from the state and the devalued naira. And as a consequence, the company stopped the operation of its DR plant effective 1992. In order to keep the rolling section of the company functioning and to remain afloat as well, DSC occasionally relied on imported billets brought in by some of its distributors which were rolled out for a fee as already explained. The irony of it all is that, DSC became dependent on imported billets which it ought to have produced locally. Under adjustment situation, too, the majority of the private steel companies closed down most of their production lines, thereby resorting to the importation of finished steel products. For instance, while CISCO wound up production since the dramatic rise in the cost of raw materials including scrapped metals for recycling following the advent of adjustment, Universal Steel Company had to import finished steel products like steel flat to keep the production of enamel wares going. Other steel stockists like the Sanusi Brothers Company Limited, located at Ibadan and Lagos equally imported more steel products like rods, bars, coils and wire meshes to fill in the gap, but the shortfall still persisted throughout the period of adjustment and even worsened after.

As regards the activity in other intermediate and capital goods-producing companies that ought to source their basic inputs from the steel companies, the performance was equally poor. It was poor in the sense that both the high import dependency ratio that characterized the structure of manufacturing and the lack of interlinks between the steel companies and the local intermediate and capital goods producers deepened. In fact, rather than increase, some measure of local capacity to fabricate few of the hitherto imported intermediate goods already attained in the country since the turn of 1970s, had declined considerably as the companies were no less technologically dependent on the metropolitan countries than they were at inception. For instance, the major foundries like the Nigeria Foundries Limited, and Roadside Engineering and Foundries Limited, both located at Ilupeju, Lagos, were still almost wholly dependent on import for their basic production inputs long after they had come on stream, and even after the steel companies had begun production. Details on the performance of the intermediate and capital goods-producing companies are spelt out in Tables 5.5 and 5.6, located in the Statistical Appendix. Because there is yet to be a broad national policy on steel development in particular, and industrialisation generally, the performance of the intermediate and capital goods industry remained unsatisfactory even after the scrapping of adjustment which the Nigerian state had introduced with the hope of helping the economy recover from the crisis.

Across both light and heavy intermediate and capital goods-manufacturing companies at the time this study was conducted, capacity utilisation was about 15 percent, one out of every two production lines was shut either temporarily or permanently, and downsizing remained on the increase. If the regimes of then Babangida and late Abacha were accused of insensitivity to the crisis in the Nigerian economy, that of Abubakar has not really been any different. As noted earlier, the crisis in which the country's economy is mired, stems basically from the nature of the Nigerian state and its mode of surplus extraction from the economy, but not the regime which has, on average, been a modality of capitalist accumulation. Hence, there is little or nothing a regime could do to tackle the root cause of the crisis without first re-orienting the state and those who manage it from using the political platform of the state to meet their parochial interests. Details on this later in the concluding chapter.

The role of the aluminium industries in the production of both intermediate and capital goods is largely supplementary to the iron and steel companies. The nature of aluminium products makes it all the more so. Cast iron, for instance, has some element of alumunium and other alloying minerals; but there is hardly an engine block which is wholly made of aluminium or steel. The aluminium industry plays the same role in the construction sector where its products are used either to complement or decorate intermediate and capital goods. It plays this role more in the manufacturing of these goods than in construction. In Nigeria, the aluminium companies import aluminium billets, flat sheets and blooms for the local fabrication of construction products like roofing sheets, nails, clads for partitioning, and frames for windows and doors. As noted earlier, the alumunium companies operating in Nigeria were all engaged in end-product activity. That is, they are involved in the downstream sector of the alumunium industry, explaining in part why they are wholly import-dependent for their raw materials. Because the aluminium industry was itself, still emerging, the foreign companies operating in that sector, were not really concerned with the issue of capacity utilisation. That did not imply it was not important; rather, it was one important issue that they planned to resolve as soon as the industry begins to flourish. The state has not really been interested in the aluminium industry too. All of this explains why the aluminium companies operating in Nigeria are merely subsidiaries of the major aluminium transnationals of the West. Little wonder they are not engaged in the extraction of bauxite and smelting, but only rolling out finished products. Even with the coming on stream of the Aluminium Smelting Company, [ALSCON] Ikot-Abasi, there is still really no attempt to engage in smelting. Rather, what has taken place, is the importation of semi-finished aluminium products for final production in the country. What Nigeria has contributed so far in the jointly-owned ALSCON, is the natural gas, which is the major energy source for firing the plant. Neither the Nigerian state, nor the Ferrostaal/Reynold of Germany and the USA respectively, which are the technical partners, had ever thought of exploiting the local deposit of bauxite in the country for the purpose of smelting it and use at ALSCON. In any case, the technical partners would not want to get involved in the extraction of the local deposit of bauxite since they were already faced with the crisis of decilining market among other problems of monopoly capitalism.

The aluminium industry only helped fill the gap created by producing construction products like frames for windows and doors, nails, tubes, and pipes for street light, and in most times, they were produced to customers' specifications. Since aluminium products were, on balance, relatively cheaper than steel products, the industry had a fairly good performance only in terms of stock turnover, but still 100 percent dependent on import for their raw materials.[12]

As for the heavy industries like the assembly plants and auto-companies which are expected to source their major inputs from the steel companies, they were 95 percent dependent on import, and in most cases, the raw materials were completely knocked down parts [CKDs]. Local assembly of some hitherto imported capital goods like cars, pickups and trucks commenced in the country in the mid-1970s, long before the DSC, the only completed integrated steel company in Nigeria, came on stream. Because industrialisation was not really on the agenda as noted earlier, the steel companies were not equipped with facilities that could produce materials that, in turn, would be used to fabricate spares for the auto-companies.

Changes in the prices of the CKDs following the devaluation of the naira since 1986, affected the production cost of the auto-companies and the price of their final products. The case of Volkswagen Nigeria Limited, located at Ojo, Lagos, brought this out clearly. Before adjustment in 1986, the locally assembled Volkswagen Beetle car, which was regarded as the people's car in Nigeria in terms of its cheapness relative to other cars like peugeot, was sold at 20,000 naira. Since the advent of adjustment, dramatic increases in the prices of the CKDs forced the price of the car to 60,000 naira in1989. By 1990, its price rose to 125,000 naira and never came down even after the official termination of the structural adjustment in 1994. Not only that, the company folded up its operation in Nigeria due largely to the poor market. In response to the general glut in the local auto-market since then, other assembly plants like Steyr, Ibadan, also closed shop. For the few that could remain in business like Peugeot Automobile Nigeria, Kaduna, it resorted to the importation of finished cars while suspending local assembly. The general response of the people to the dramatic rise in price of cars, was the large scale importation of fairly used cars from Europe and America, thereby deepening the crisis in the country's auto industry. For, compared with new cars, the fairly used ones were and are still both cheaper and within the purchasing power of more Nigerians. So far, the performance of Nigeria's intermediate and capital goods-producing companies remained generally poor, characterized by heavy undercapacity utilisation. Let us now turn to the situation in Korea.

5.3.2 *South Korea*

Generally, the performance of South Korea's steel industry was better than Nigeria's both in terms of capacity utilisation and the acquisition of tremendous local technological capability to manufacture a wide range of products that are themselves, intermediate and capital goods, and serve as inputs for producing these capital goods. Unlike in Nigeria, too, where both the steel companies and other capital goods-producers were marred by undercapacity utilisation and virtually dependent on imports for their basic production inputs, the situation was quite different in South Korea. In Korea, most of the private steel companies and other firms that were engaged in the production of intermediate and capital goods, were about 80 percent dependent on POSCO for their basic raw materials with the remaining 20 percent sourced through imports.

For its parts, POSCO had established chains of end-user companies which, in turn, manufacture intermediate and capital goods locally, unlike in Nigeria, where the state-owned steel companies, namely the DSC and the three steel rolling mills at Katsina, Jos and Oshogbo, hardly operated for a total of 365 days since their inception in 1982 much more talking setting up similar end-product companies. Furthermore, POSCO had its own internal capability to produce finished capital goods. As for the heavy industrial sector of the Korean economy, its activity was dominated by the big business groups, *chaebols* , with the POSCO's major subsidiaries and a few other local private companies accounting for the rest. Planned by the state to depend largely on POSCO for their major raw materials while their import dependency ratio would be reduced over a period, the intermediate and capital goods producers were all engaged in manufacturing a wide range of hitherto imported goods such as household utensils, railwagons, tracks, shoes and brakes, storage silos, ships, containers, automobiles, heavy machineries and their spares. All this, is by no means an assumption that, there was no shortfall in meeting the installed capacities of the various production lines of the steel companies and other intermediate and capital goods producers. As it shall soon be disclosed, there were shortfalls.

Between 1973 and 1983, the four phases of POSCO's Steel Works located at Pohang with an annual installed capacity to produce 11 million metric tonnes of crude steel, were completed and commissioned. Within this period, too, an estimated total output of about 90 million metric tonnes of crude steel steel was planned to be produced. Out of this, a total of 50 million tonnes of crude steel was produced in the period under review. In terms of performance therefore, it represented 55 percent of its cummulative installed capacities over 10 years. With the construction and completion of the four phases of Kwanyang Steel Works between 1985 and 1992, POSCO's second and larger integrated steel plant, a total of 30 million metric tonnes of crude steel was additionally produced. Together, POSCO had an installed capacity to produce 41 million metric tonnes of crude steel yearly. Between 1983 and 1992, POSCO produced 150 million metric tons of crude steel, and by 1995, produced 230 million metric tonnes which made the company the second single largest world steel producer after Nippon Steel of Japan for that year.[14]

If POSCO's total crude steel of 41 million metric tonnes is added to the estimated total 15 million metric tons of crude steel from the private companies, then, Korea has an internal capacity to produce an average of 56 million metric tonnes of crude steel yearly.[13] In fact, since 1990, the overall performance of the Korean steel industry in terms of crude steel production has been quite impressive. For instance, from an SSR of 80 percent in 1990, it rose to 95 percent in 1996. Table 5.7, located in the Statistical Appendix vividly presents the trend of growth in the production of crude steel in Korea for the period under review.

As noted earlier, Korea's steel industry was heavily dominated by the state, that is, POSCO and its subsidiaries. This is about one feature which is common to both the steel industries in Nigeria and Korea. In terms of performance, unlike in Nigeria where the public steel companies were largely a failure, about 90 percent of the total crude steel produced in Korea, came from POSCO, with the remaining 10 percent accounted for by the private steel companies. One of the reasons adduced for the low percentage share of the private steel companies in the total crude steel produced in Korea, stemmed largely from the state's industrial policy, but not that they could not really perform. As pointed out earlier, the Korean state made it clear in its national industrial policy that, POSCO was set up to act as the launching pad for the country's industrial and technological age. Within this broad industrial policy, the big business groups were to be promoters of the country's export-led industrialiastion policy, while also meeting the local needs of Koreans for other intermediate and capital goods. Therefore, the state did not pay much attention to the private steel companies virtually all of which were owned by indigenous investors. What the government did instead, was to ensure that POSCO met a lion share of the basic raw material needs of the private steel companies while they source the remainder locally through the recycling of metal scraps. Hence, their performance in this regards was rather low, but still relatively better than their counterparts in Nigeria. It should be noted that, the poor performance of Korean private steel companies in comparison with those in the public sector, was rooted in the industrial policy of the Korean state, which has already been explained. This is one of the major contradictions in the industrial policy of the Korean state.

With respect to the production of finished steel products and other intermediate and capital goods, POSCO was by far, a better performer in compariosn with Nigeria's integrated steel companies. Unlike in Nigeria where the product mix of both public and private steel companies was basically constructional steel products like rods and wires, that of Korea's POSCO and the private steel companies was of a wider range. In terms of the general performance of the country's steel industries in the production and consumption of finished steel products, there was an increase since the 1980s which continued into the 1990s. Table 5.8 located in the Statistical Appendix, illustrates this trend of impressive performance for the period, 1994-1996. As for the specific details on POSCO's performance, both its production lines as already shown in Table 4.2, and its manufacturing of finished products all witnessed a significant increase in capacity utilisation. For instance, the hot rolling mill produced not only at an average of 90 percent of its installed capacity, but also, accounted for 50 percent of all the finished products rolled out of the company. Furthermore, its cold rolling mill accounted for 30 percent of the finished products, while 10 percent each was produced from the plate making section and the stainless steel production lines respectively.[15]

Since 1994, POSCO has gradually come close to the peak of its capacity in the production of crude steel. Perhaps, this explained why the projection for crude steel production had remained fairly constant between 1995 and 1997, while other sections of the various plants had witnessed some increase and also performed very well. The only exception to the overall impressive performance in the company's production history, was the rolling line for the production of reinforced steel bars. On account of some shortfall in the domestic demand for the product by the construction industry in the period, 1996-1997, there was a decline in the production of steel bars. According to both field and office staff of POSCO who the author spoke to at Seoul, Pohang and Kwanyang in 1997, the shortfall was temporary and would soon be overcome. This is all the more so as an average of 56 percent of the steel products was still consumed by the construction industry. However, with the re-emergence of the structural crisis in the Korean economy in late 1997, the first having occurred in the late 1970s, and attempts to resolve it failed to deal decisively with the root cause of the crisis, most of the country's industries including the steel companies, were forced by both the state and the IMF to downsize and operate in line with the market mechanism. In other words, POSCO and the private steel companies were off to a more serious crisis.

As for the production of stainless steel, POSCO did not perform well relative to other products. Why? In part, it was because of the glut in the global steel market since 1975 which also accounted for the low domestic demand for the product in Korea. That is not all, as the company had established the production line for stainless steel ostensibly to meet local demand for the product, and having partly met that objective, it wanted to maintain skeletal production until the world steel market improves. In fact, most of the steel producing countries in Europe like France, and even Japan, have maintained a low profile in the production of stainless steel until the cessation of the crisis in the global steel market.

As for the private steel companies, some of the notable ones are already presented in Table 3.4, their performance was better in comparison with Nigeria, but not too well in the context of the overall Korean steel industry. While the private steel companies operated at about 15 percent of their installed capacity, representing 5 percent of the total steel output in the country, the situation in Korea was quite different. Though they equally accounted for no more than 20 percent of the total steel output of Korea, the private steel companies were better organised and able to produce locally, some of the hitherto imported intermediate and capital goods. By the industrial and steel policies of the Korean state, the majority of the private steel companies in the country were more or less made captive to the heavy industries. Also, they had their production history indirectly tied to the ability of POSCO to supply their production inputs most of which were semi- and finished steel products. In essence, the spate of growth of Korea's private steel sector was to a large extent, intricately tied to POSCO's ability to deliver.

Not only that. The Korean government had it, as an unwritten industrial policy, to prohibit any private steel company from intervening into its upstream sector. What this means in practice is that, the government would hardly guarantee any loan among other forms of requests from private steel companies negating this policy stance. For instance, the Korean government would not support any proposal from a private steel company to import finished steel products which are produced by POSCO since the company was already planned to meet their local steel needs. Officials of POSRI told the author in Seoul in June 1997, that one of the major reasons for the collaspe of Hanbo Steel was its violation of the state industrial/steel policy by intervening into the upstream sector of the country's steel industry.[16]

But among the private steel companies, the thinking was not the same. According to the officials of Korea Steel Association [KOSA] who the author interviewed in Seoul in 1997, the Korean government was largely to be blamed for the general trend of undercapacity in Korea's private steel sector. The poor performance across the private steel companies was because POSCO was alledged to have always favoured the *chaebols* to the detriment of their member-companies. Recycling of metal scraps constituted an insignificant portion of the ability of the member-companies to augment production. So, they could not do much to improve the performance of the steel companies.[17]

Hence, there was an average of 25 percent capacity utilisation across the private steel companies like Inchon Iron and Steel Company, Donguk-Steel Mill, Kangwon Industries, Hankook Steel and Mill Company and Sammi Steel Company. Table 3.4 had already shown a list of some of the major private steel companies in Korea. In addition to Hanbo Steel Company that became bankrupt in March 1997, the financial crisis that rocked the KIA Group of Companies seriously affected one of its subsidiaries, the Kia Steel Company located at Kusan. As of 1997, the company operated at 10 percent of its installed capacity.

In terms of POSCO's production of semi- and finished steel products as inputs for its end-user companies, the company performed well in comparison with Nigeria's DSC and the steel rolling mills. Table 5.9, located in the Statistical Appendix, clearly explains in a comparative context, the nature of steel consumption both by product and industry in Nigeria and South Korea for the period, 1980-1995. As noted earlier, the nature of steel products produced and its pattern of consumption, all give an insight into a country's level of industrialisation, particularly its capacity to manufacture. As can be seen in Table 5.9 for instance, while the total output of finished steel products in Nigeria for 15 years was 328,000 tonnes, Korea produced 70.72 million tonnes within the same period.

With respect to the consumption of finished steel products by industry, both Nigeria and Korea still had more than half of its total output consumed by the construction industry, an indication that their steel industries were still developing, though they were at different stages of development at the time this study was conducted. As explained in Figure 5.1, while Nigeria was still at the 'take-off' stage, Korea was already at the stage of 'late growth' in steel development. The difference in the stages of steel development of both countries has been brought out mostly clearly in the pattern of steel consumption by industry as shown in Table 5.9. Rather than increase, the share of overall steel consumption by the construction industry in Korea had fallen from 57 percent in 1980 to 47.1 percent in 1975. That is, a 10 percent reduction was achieved within a period of 15 years. In Nigeria, it was all an increase in the share of construction of the total output of finished steel products for the same period.

As explained earlier, manufacturing of intermediate and capital goods did not really take place in Nigeria in the strict sense of using iron ore to produce iron and steel and, in turn, used for the final production of these goods. Therefore, what the 3 and 2 percentage shares of the steel consumed by the automobile companies in Nigeria in 1980-85 and 1995 represented respectively, were no less than the importation of CKDs, and the local fabrication of spare parts using scrapped metals. For, the steel companies in Nigeria had, at the time this study was undertaken, no production components for rolling out flat steel products, the major input for the production of the majority of the intermediate and capital goods. In Korea, the contrast was quite distinct both in terms of the percentage share of finished steel product consumed by the auto-industries and in the real sense of manufacturing locally, most of the hitherto imported intermediate and capital goods. For instance, while 5.3 percent of the total output of finished steel product was consumed by the automobile-producing companies in 1980, the figure increased to 12.3 percent in 1995, representing about 60 percent increase.

With respect to fabrication, there was a decline in the percentage share of steel consumption from 23.6 in 1980, to 15.5 percent in 1995. That was not really indicative of poor performance in the industry; rather, it was due to the re-allocation of consumption of finished steel products from fabrication industries to the actual production of machineries whose share of consumed finished steel product shut up from 4.4 to 8.3 percentage points. See Table 5.9 for details. In fact, it was as a result of increased activities in the production of inputs for the manufacturing of intermediate and capital goods that made the major end-user companies to perform relatively well. For instance, Korea's auto-industry ranked the fifth in the world automobile industry; Korea's shipbuilding industry was also the second in the world's shipbuilding industry; Korea became the third builder of containers in the world. Impressive as the performance of the Korean steel industry was, it still contrasted sharply with the situation of Japan especially in terms of the structure of steel consumption by industry. In Japan, the percentage share of the construction industry of the total output of finished steel consumed, was 43.3 in 1994, while manufacturing took 56.7 in the period under review. This trend of more steel consumption by manufacturing industry has brought out very clearly, our earlier explanation that, the more industrialised a country is, the less steel is consumed by the its construction industry. Table 5.10, located in the Statistical Appendix, sheds more light on Japan whose experience with steel development had contrasted sharply with the situations in Nigeria and South Korea.

One of the major hallmarks of POSCO's performance during the period under review, was its capability to transform most of its components into full-fledged companies, though they were still wholly public. It is these hitherto various production units-turned subsidiaries of POSCO, that helped in the local production of intermediate and capital goods relying exclusively on their parent company for most of their basic raw materials. POSCO's subsidiaries, namely, POSMEC, POS-M, POSREC, POSEC and POS-Energy, have come to constitute a major aspect of Korea's capital goods industry not really as a national policy, but by the pragmatic approach of the Korean government to both steel and industrial development. The officials of the Korean government the author spoke to, were of the belief that, following the completion of POSCO's two major steel works located at Pohang and Kwangyang, it was better to transform the various major units into state-owned companies ostensibly to complement the efforts of the *chaebols* in the local production of intermediate and capital goods in the country instead of establishing completely new heavy industries.

At the time this study was conducted, these specialised companies were well developed to the point that they began to export some aspect of steel technology to some South East Asian countries like Vietnam by way of helping them to design, build, install and repair various major components of the integrated steel plant, mini- and rolling mill. For instance, POSCO Refractories Company Limited, POSREC, first of all the specialist companies to be created, began in 1972, as the internal unit of the company that produced materials for the maintenance of its coke oven batteries and the blast furnace. Later, it started producing refractory materials in large scale both for internal use and as export to emergent steel-producing Asian countries like Vietnam. The same pragmatic approach of the Korean state to self-sufficiency in steel, also informed the establishment in 1982, of POSCO Engineering and Construction Company Limited, POSEC, particularly after the completion of the final phase of Pohang's Steel Works. The POSEC helped a lot in the overall construction of Kwangyang Steel Works of POSCO. As an engineering and construction company, POSEC's main activity was to build steel works, especially parts of the blast furnace, direct reduction plants and mini-mills, complex bridges, ports and sky-scrappers. POSCO Machinery and Engineering Company Limited, POSMEC, which started also in 1982, has as its major activity, the designing, manufacturing, installation and maintenance of steel making machines, conveyor steel rollers and their spares. Although these companies carried out rudimentary internal maintenance services, it was the POSCO Maintenace Company Limited, POS-M, established in 1987, that undertook major repair services for all POSCO's companies and for other steel plants in the country. POSCO Energy, POS-Energy, was established to design, produce parts of, install and operate thermo-electric power plants, and the development of energy resources for both POSCO and other heavy industries in Korea. Locate Table 5.11 in the Statistical Appendix for details.

One important lesson that the Nigerian state has to learn from South Korea state approach to steel development generally, and self-sufficiency in steel in particular, in the context of POSCO's subsidiaries, is that POSCO only thought of branching into specialised services in the steel industry only after its major steel works were completed. In other words, it was not a way out of a protracted crisis over steel development as was the case in Nigeria where government had attempted to commercialise the completed light sections of the Ajaokuta steel, but without success. Worst still, the construction of the company's blast furnace, which began since 1979 was still uncompleted as of May 1999, whereas it took three years to get the first phase of Pohang Steel Works that is ten times larger than Ajaokuta steel plant, completed. To commercialise the light sections and foundry of the uncompleted Ajaokuta steel company with its blast furnace still under construction is no less than an admission of failure on the part of the Nigerian state. More details on this later.

Like the *chaebols*, the custodians of the Korean state were of the belief that POSCO could also help promote its export-oriented industrialisation policy in particular, and the country's international economic relations in general especially during the cold war period and after. How POSCO's subsidiaries helped accomplish Korea's export-oriented industrialisation has been partly explained in the activities of these specialised companies as shown in Table 5.12 located in the Statistical Appendix. It is equally more important to note that, soon after the completion of the company, those in charge of the Korean state were of the view that, one of the ways to industrialise Korea given the global political economy of industrialisation, was to reduce drastically, if not to halt, the country's technological dependence on Japan in particular and the West. For instance, rather than having its steel companies tied in a dependency relationship with Voest Alpine of Austria in particular, and the steel industry of the West in general, which installed POSCO's concast plants, the government was of the view that it was better to establish a Korean company to undertake the task of future repairs of the steel plants.

As explained earlier, this was one of the main reasons that informed the establishment of POSMEC. With a reasonable level of perfection in this aspect of steel technology, and coupled with the role of Korea in the politics of the cold war in the East Asian region, it became rather easy for the Republic to export its relatively newly acquired expertise in steel development to some of the East Asian countries like Vietnam. The choice of Vietnam among other countries by the Korean government, dates back to the internal political crisis in the country in the 1970s in which the Korean government contributed troops to support the cause of the government of the US in that country. For instance, POSCO had been supplying long and round steel products to Vietnam since the 1970s before it finally signed a joint venture agreement with VSP, Haiphong in 1994. Through this agreement, a steel rolling mill, named as VSC-POSCO Steel Corporation with 35% POSCO, 34% VSC, 10% Haiphong Engineering, and 5% POSTADE [a branch of POSCO], was established to roll out steel bars and wires for sale in Vietnam. The company was capitalised with US$16.8 million. In Vietnam, too, POSCO had another joint venture agreement, but this time with SSC F, located at Ho Chi Minh. With an equity structure of 50% POSCO, and 50% SSC F of Vietnam, and named as POSVINA Company Limited, it produced galvanized steel products for sale in Vietnam.

Largely on account of political reasons too, POSCO entered into joint venture agreement with the United States Steel to establish USS-POSCO on a ratio of 1:1. In other words, both companies had 50% each in the US$388million equity of the company. The Pittsburg-based USS-POSCO Industries, UPI, which was a transnational marriage between the two steel companies, was established to produce and sell cold-rolled steel products in the US, and with the hope of also exporting some the steel products to Latin America. POSCO also established the Pohang Steel America Corporation, POSAM, in 1984. With an equity of US$246.8million, POSAM was wholly owned by POSCO. Its major objective was to help promote the sale of POSCO's steel products in the US. See Table 5.12 for details.

As pointed out earlier, the US steel industry, unlike its counterpart in Western Europe, did very little to shape the global steel industry. Inspite of the fact that the then USSR was about the world largest producer of crude steel, it was not part of the global steel industry. And as a consequence, it hardly influenced the steel industry of the global market economies. One of the major reasons for the insignificant influence of the American steel sector on the global steel industry, was informed by the inward-looking steel policy of the country's government which made the steel companies to be basically concerned with the satisfaction of the US local steel demand. With the American steel industry entering the stage of market maturity in the late 1960s, most of the steel companies could not remain in business as glut in the local steel market had set in, precipitating the closure of most of production among other problems. In essence, the receeding American steel industry could be taken advantage of by some of Asian newly industrialising countries like South Korea. Japan had already cashed in on the American steel industry. The fact that the US used South Korea as bulwark to stem communism in East Asia during the cold war, made it easier for POSCO to enter into the American steel market either through joint venture or setting up market outlets for its steel products.

Within the South East Asian region since the turn of the 1970s, Korea's POSCO was next to Nippon Steel of Japan in terms of control of the sub-region's steel market. This enabled POSCO to take advantage of the huge untapped steel market of South East Asia by setting up rolling and mini-steel mills in the sub-region as a means of promoting the country's international economic relations there. Largely due to political reasons, Nigeria and the rest of Africa, still did not really factor into Korea's external trade/steel policy as of the time of this study. It is not that a pro-Nigeria/Africa and Korea's external trade/steel policy would have really done much to change the underdeveloped steel industry in Africa if its post-colonial state refuses to have industrialisation on the agenda. Rather, what should be of interest to Nigeria in paricular and Africa generally, is how Nigeria could be made to play similar role in West Africa as South Korea is doing in the ASEAN.

5.4 Concluding Remarks

It is clear from the above analysis that the performance of both the steel companies and the intermediate and capital goods-producing companies in Nigeria was generally heavily marred by undercapacity utilisation, and closure of more production lines and shops. This contrasted sharply with the performance of the Korean steel industries and captial goods producers which recorded an impressive growth. The steel industries of both countries were however, not without problems; problems which have bedevilled their operations at their various levels of development. In Nigeria, the problems facing its steel industry are still basically political, though they tend to present themselves as economic and technological. Like in Nigeria, part of the problem of the Korean steel industry is political. Nature did not endow Korea with most of the basic steel raw materials, and this is a serious problem compared to Nigeria where they abound. In the light of the differences in the nature and kind of problems which the steel industries of Nigeria and South Korea were faced with, the next chapter critically analyses in a comparative perspective, not only how these problems have come to be, and constrained their operations, but also, the prospects of the steel industries of both countries.

NOTES AND REFERENCES

1. The International Iron and Steel Institute, Brussels, Belgium, set the standard at 100 percent. See also, *The Statistical Steel Yearbook* of the IISI, 1990.
2. *ibid.*
3. *ibid.*
4. *ibid.*
5. *ibid.* See the 1995 and 1996 issues of *The Statistical Steel Yearbook*.
6. See Daniel Omoweh, 'The Nigerian Steel Sector in the Global Steel Industry', *Annals, op. cit.*
7. Based on the author's field trips to DSC in 1989, 1990, 1993, 1995, and 1996.
8. *ibid.*
9. According to the management of the company, there was every tendency for the federal government to direct the loan to uses instead of the main purpose for which it was meant if it were paid in cash. It was for this reason that the company's management agreed with the Kobe Steel to convert it into billets. In any case, the company had no option since Kobe Steel itself, wanted the loan in the form of billets.
10. As of May 1995 when the author visited ZISCO, the company was already shut down for over five years. However, the Mugabe administration was able to source external fund with which the repair of the collasped flat steel production line and the entire steel company began in 1997.
11. Extracted from *Annual Abstract of Statistics*, Lagos, Federal Office of Statistics, 1987, 1990, 1995, 1997 and 1998.
12. In fact, within the three months of trial production in 1998, the technical partners exhausted the commissioning raw materials for ALSON hoping that the Nigerian government would keep to its own part of the agreement to fund the company. But theNigerian

government said that it has no other money for the project.The ALSCON is already cash strapped in less than six months of its take-off. Officials of the Aluminium Department of the Ministry of Power and Steel could not explain why government failed to provide fund for the running of the company.

13. Data were extracted from POSCO's production departments.
14. Data were extracted from KOSA's statistical documents on steel production of member-companies.
15. Data were abstracted from POSCO's production department.
16. Based on author's checks in 1997
17. *ibid.*

Chapter Six

Problems and Prospects

6.1 Introduction

As its primary object, this chapter identifies and analyses in a comparative perspective, the problems and prospects of the steel industries of both Nigeria and South Korea. In terms of the problems, it is undertaken in the contexts of the factors militating against the operations of the steel companies, the factors militating against their contribution to the production of intermediate and capital goods, and overall linkages with the rest of the economy. As for the prospect of the steel industry, its future is examined in the context of the potentials of the sector to exist inspite of the problems. The discussion begins with the problems.

6.2 Problems

To start with, the method of analysis adopted in this study is such that, it simultaneously brings into limelight, both the nature of the problems and the prospects of the steel industries of Nigeria and South Korea. All the same, attempt will be made here to pinpoint the major ones varied as they are. Though the problems present themselves generally as economic and technological, manifesting in the form of financial squeeze, crisis in management techniques, poor intersectoral linkages, overinvestment, acute raw material shortages, they are basically rooted in politics. As a political problem, the main issue is the state, its nature of surplus extraction and conception of industrialisation generally, and the steel development in particular. And unless the root cause of the political problem is resolved, the techno-economic problems which are no less than the symptoms of the latter, would be difficult to deal with. Hence, the analysis of the problem starts with the state.

[i] The State: To begin with, the problem is not that the state made a mistake by intervening in steel development in both Nigeria and South Korea. For, the past experiences of the major steel producing countries in the world today, like Germany, France and Japan, the state had led the development of their steel sectors in the past before it rolled back after a significant level of growth was achieved. Rather, the real problem is the nature and politics of the state, its mode of capitalism and conception of industrialisation and steel development. Let us explain how all this constitutes a problem to the development of steel industry of both Nigeria and South Korea, beginning with the former.

In Nigeria, the post-colonial state is not only all powerful and everywhere, but also, still largely remains an extraction of foreign capital instituted to further the material interest of its custodians. What became of utmost concern to those entrusted with the management of the state as noted earlier, was to inherit, but not to change the exploitative policies and structures of the colonial state. It was partly the nature of the Nigerian politics, especially how it became a mode of accumulation that explained why industrialisation and steel development failed to really interest the Nigerian state. If for anything, the state-owned steel companies, among other public companies, only provided avenues that helped the political class, the state managers and their allies to accumulate. And this was to frame to significant degree, the essence of the Nigerian politics: a means to amass wealth. Hence, warfare as opposed to dialogue, became the only currency of Nigerian politics. Not surprisingly, those who have captured state power want to maintain it by means and at all cost including denying the losers of their inalienable rights. All this explained why the Nigerian state was and is still not only repressive, but the grim struggle between and among the institutional groups that make the state a reality, to use state power to accumulate pose fundamental problems to the country's quest for industrialization inclusive of steel.

Under this kind of atmosphere of bloody political competition coupled with the unbriddled quest by the state custodians to accumulate using political power as against entrepreneurship, it is natural to expect that little or no consideration would be given to the capacity of the country's economy to reproduce itself generally. What the state has done so far, is to use its policies and actions to further render the local economy as an appendage of the metroplitan countries. This guarantees the accumulative base of the custodians of the state and their cronies, local and foreign inclusive of foreign capital.

Little wonder the state was hardly interested with the development of the local productive forces in the economy in order to wrest its control from foreign capital, explaining why its capitalism has, on account of its penchant for rent, connived with foreign capital to undermine the capacity of the local economy to reproduce itself. And as a consequence, state-owned big companies like DSC, which ought to act as a 'spoke' rotating the 'hub' of the country's industrialisation, not only become diversionary, but also, was transformed into a microcosm of the state. A microcosm of the state in the sense that it becomes a platform where the intense battle rages on for the control of political power and accumulation between and among the various institutional groups that make the state a whole and reality. For instance, the public steel companies became part of the diversions in the sense that, the nature of state politics has derailed them right from inception so much so that they could not be used to quicken Nigeria's industrialisation. If for anything, they are turned into fertile grounds for the state to make plum but unproductive political appointments to its members. Which is why the singsong of the Nigerian state since 1979 that 'steel is the bedrock of Nigeria's industrialisation' is no less than a second catch phrase being used by the political class to solicit the political support of the people, the first being the clarion call by the nationalist movements for political independence. For, after the collaspe of British colonial rule in 1960, the political class turned to the adoption of a defensive radical political posture that looks as if its members are on the side of the struggle for economic development with people whereas its real interest still remains with foreign capital. In essence, steel development has, from inception, lacked all the seriousness that it needed to effectively take-off and survive. This is partly how the Nigerian state, its politics and method of accumulation have constituted formidable constraints to Nigeria's quest for industrialisation. Further analysis of how state politics dwarfed the steel sector will be helpful at this point.

Before the public steel companies were constructed between 1979 and 1983, the Nigerian government had, in collaboration with foreign experts of iron and steel, conducted techno-economic studies. Unfortunately, rather than these studies guide the establishement of the state-owned iron and steel companies, it was politics. For instance, in its choice of technological route to be taken for iron and steel making as well as product mix, the state managers did not consider it wise to have the local iron ore ready for use by the steel companies while they were being built. It was after the DSC had come on stream that the some cosmetic efforts were made by the government to develop the Itakpe iron ore deposit. All this was compounded by the fact that, as of 1979, when the Nigerian government embarked on steel development, there was no industrial policy; and even as of the time this study was conducted, there was still none in existence. In essence, the public steel companies were established without a policy guide, partly explaining why they were beset with confused and conflicting interests right from their inception. Partly in an attempt to reproduce its means of survival, steel development was only embarked upon by the state custodians as a measure of recycling part of the oil wealth. They were constructed on the same framework that informed the state's philosophy of developing real estate, in which its ultimate interest is to lease the property for rent. It is in the same context the public steel companies were planned to be rented or leased out to both local and foreign capitals to manage for rent. In fact, General Obasanjo whose regime had, in 1979, signed the contracts for the construction of the five public steel companies, had one of his companies involved in a business deal with DSC and Oshogbo Steel Rolling Company. In case of DSC, his company had taken advantage of the financial squeeze facing DSC and imported iron ore for it to use to produce steel rods and coils for which a fee was paid. Despite the fact that his company paid the fee, DSC sold the rolled products. The matter was yet to be resolved when he, General Obasanjo that was framed up in coup plot to overthrow the late Abacha's junta, and he was subsequently jailed. The issue at stake is not that, it was a company owned by General Obasanjo was involved in the deal; rather, that, in part, is a demonstration of the ultimate aim of the state for establishing the public steel companies. Under this frame, Nigeria's quest for industrialization will, for a time to come, remain elusive. The contradictions unfolded more in the provision of raw materials.

As for the basic steel raw materials like coal, the initial plan was to mine it for export. For iron ore, the plan of government was to develop the Itakpe iron ore deposit for use at Ajaokuta, while DSC was to depend on import despite the relatively ferrous grade of the Agbaja iron ore reserve which was left unexploited. Not even an estimated $2.5 billion already sunk into the development of steel raw materials yet DSC was still reliant on import for iron ore would force the state into a rethink. It was largely on the account of non-commitment of the state to the steel project that explains why soon after take-off, DSC, and the three inland steel rolling mills located at Katsina, Jos and Oshogbo were mired in protracted financial crisis, deepening dependence on imported raw materials that even abound locally, indefinite closure of production lines until they were shut down, overinvestment and conflicting management styles.

These were some of the basic contradictions arising from the political economy of the state's approach to industrial accumulation. The delapidated condition in which the public steel companies were at the time of this writing, was indeed, what the Nigerian state wanted them to be, and not they were really faced with unsurmountable problems. For one thing, the poor state of the companies creates an avenue for request for more funds for their repairs, which were then misappropriated by the state managers. That is not all. The decision to commission the DSC with imported iron ore was political because it profited a tiny group within the state to do so. So, too, was the financial squeeze that all the state-owned steel companies were faced with, political but not economic as the government would want us to believe. In essence, if the public steel companies have to really recover from the crisis they have been faced with since their inception, then the starting point is to tackle the politics of the Nigerian state and its model of capitalism. More details on this in the next chapter. Let us now turn briefly to the relationship between the state and the private steel companies.

Since the Nigerian state embarked on steel development without both steel and industrial policies, then it is natural to expect that what goes in the private steel sector will hardly interest the Nigerian government. As pointed out earlier, one of the reasons why the state intervened in the steel industry, was the inability of the private steel companies to cope with the domestic steel demand in the country. But with the coming on stream of the public steel companies, the situation still did not really change for the better as domestic steel need continues to be largely met through import. One reason why it has been so, is not unconnected with the absence of an industrial/steel policy that ought to have regulated Nigeria's steel industry, particularly the kind of steel products being imported into the country and for how long. As it shall soon be explained, the role of POSCO in guiding the Korean steel industry in this respect, is quite instructive for Nigeria. Let us expantiate more on the problem caused by the absence of steel policy.

Given the initial huge capital outlay required in establishing the DSC, the only integrated and functional steel plant in Nigeria until its indefinite closure in 1996, the Nigerian government ought to have used it to structure the country's steel industry in general, and in particular, the operations of both public and private steel rolling mills in terms of the kind of steel products they produce, importation of steel products, and the steel technology. Because the state, is itself, not really interested in steel development as noted earlier, both the DSC and other steel mini-mills have the same product mix, an indication of a fundamental contradiction in the path that the state took to steel development. Not surprisingly, both the public and private steel companies operated independent of each other, whereas they are supposed to complement the activities of each other in order for them to collectively promote the country's industrial growth. In fact, the private steel companies were, at the time this study was conducted, more engaged in the importation of finished steel products than producing them locally, making the products of the public steel companies relatively more expensive than the imported ones in the country's local steel market. For instance, a ton of steel rods produced by the public steel companies was 30,000 naira compared with 22,000 naira for a ton of imported steel rods as of 1987. The trend of more expensive locally made steel products continued into the 1990s as some government officials were also indirectly involved in the importation of steel products. Hence, the government could not check the dumping of basic finished steel products like rods and coils in the country. Let us turn to the discussion on South Korea.

In South Korea, the nature of the problem posed by the state in the country's steel industry was quite different from Nigeria's. However, both countries still have common problems in some areas, which will be highlighted in the course of the discussion.

Between 1945 and 1961, the nature of the post-colonial economy of South Korea was largely agrarian, aid-dependent, consumerist and with a weak and pliant state. Locating the root cause of the problem that Korea was faced with in 1961, in purely economic context, whereas it was indeed, political, was one of the major problems of the Korean state. For, Korea's economic underdevelopment was largely rooted in the politics of the Korean state in the periods, 1945- 1948 under the US Trusteeship, and 1948-1961 under Sygman Rhee. Even the decision to reconstruct the Korean state after 1961 was still political despite the awowed commitment of the Park regime to move the Korean economy in the direction of growth. In essence, those entrusted with the management of the post-1961 Korean state were all products of the political process of re-inventing the state in South Korea. So, it is in the politics of the reconstruction of the Korean state that the problem it later posed to the industrialisation project in particular, and the path it took to the overall development of the economy generally, is to be located. For instance, with the military in power, the Korean state was authoritarian. This was all the more so as the members of the military weilded much influence as they constituted the hegemonic class of the Korean state structure. The Korean state had slightly the same experience with Nigeria's, where the state was and still authoritarian. However, while authoritarianism was used to bring about some level of economic and industrial growth in Korea, the opposite was the case in Nigeria. One major explanation for the sad case of Nigeria was that, the military was first seen as a vanguard of the ruling oligarchies and the members of the political class who lost out in the struggle for state power, until since the early 1980s, when it bacame a formidable political class of its own, holding on to political power by all means and at all costs. Indeed, the military is no less than a mode for amassing wealth illegally in Nigeria. Based on the experience of the past 35 years of military rule in Nigeria, the military came to the political scene only to consume public wealth notably the proceeds from oil export and production, but not to produce it.

But authoritarianism also had its limitations particularly as it affected the efforts of the Korean state to bring about a balanced economic development of the Republic. The development policy of the Korean government since the days of then General Park had had the state at its heart, and still has not changed as of 1998 under the regime of Kim Dae-Jung. The Korean state dominated the economy and acted as a market on itself in most cases since 1961 till 1998, and with little or no hope of having the trend changed. A lot of the capital injected into the economy were public, and it came in largely as foreign loans to the Korean government. All this explained why the private sector was still relatively undeveloped and contributed very little to capital formation in the Korean economy. In other words, rather than been propelled by market forces, the Korean economy was still led by the state. The argument advanced by the officials of Korean Development Institute that, the policy and action of the Korean government in this respect, were meant to protect its fragile economy at that time, was no longer tenable as it was still continued with as of 1998. In fact, the inability of the state to decide when to roll back from the heart of the Korean economy is a major contradiction in the national industrial policy of the Korean government. How did all this pose a problem to the country's industrialisation project?

As explained earlier, the Korea's political system is quite far from being democratic; yet, it tried to mimic the basic features of democracy by having elections and installing the structures of governance and rule of law. In terms of its economic policies, inspite of all pretensions to make the economy look as if it was driven by the forces of the market, the reality is that, it is still controlled by the state, operating a more or less a centrally-planned economy. For instance, in theory, the Korean government had, since then regime of General Park adopted monetarist pricinples, but in practice, the state still called the final shot in terms of national economic policy guide, project initiation and funding, and implementation through the *chaebols*, while leaving the economy still closed to foreign capital.

From all appearances, Korea's industrialisation project was the brainchild of the Korean state. And given its nature and the materialist interests of those managing it, the state could not really decide at what point it should stop acting as an investor and begin to regulate the industrial activities in the economy. Its powerful bureaucracy that was formed to withstand the pressure from the opposition, made it all the more so for the state's reluctance to delink from the economy. Little wonder the Korean state was better described authoritarian-bureaucratic. That is not all as the path which the Korean government took to industrialisation and steel development was not people-oriented. In fact, the industrial structure of the Korean economy in the period before 1961 hardly entered into the planning of the country's industrialisation since the regime of General Park. This was clearly demonstrated in the state's mode of exploitation in the economy as evidenced in its questionable role in establishing and funding of the *chaebols*, and its long-standing anti-labour laws. Since the industrialisation project of the Korean state turned out to be a haven for the material sustanance of the political class, particularly for its members who are in government, the state will continue to be hesistant to roll back from the mainstream of the economy. All this partly explains why the Korean government still could not really open up its economy to the exports and capital inflow from the member-countries of the OECD long after joining the Organisation. Why? Given the brittle foundation of the Korean economy, opening up its economy to the members of the Organisation would result in its capitulation. Which is why many scholars of the Asian economies are still at a loss about the basis of the wide publicity the Asian NICs received from the World Bank survey of 1993. That is not all as Korea's economy was still not marketized in the sense of what obtains in the West. All this gives an insight into the real causes and nature of the current crisis in the Korean economy since 1997.

With respect to the steel industry, though the Korean government had made significant progress in comparison with Nigeria, its approach to steel development was not without problems. There are contradictions in the steel policy of the Korean government. Through POSCO, the government was able to dominate the country's overall steel industry to the detriment of private capital. Because the Korean state was all powerful, its POSCO still remained in control of the country's steel sector, particularly in its upstream which the company had a monopoly. Obviously, the current trend of state's dominance of the country's steel industry portends difficulty for the Korean steel industry to attain market maturity since it is the private steel companies that ought to lead a country's steel sector into that stage of growth. Worst still, the Korean government wanted to make POSCO look like a private company and even a steel transnational, whereas it is nothing of the sort. The equity structure of POSCO for instance, was still 95 percent owned or directed by the Korean state and its agencies like the Korean Development Bank.

As noted earlier, the Korean economy was not only growth-obsessed, but suffered from stability and equity. It was within that context that POSCO had operated, making it all the more difficult to really compare the performance of the company with its counterpart in Japan, Germany and France. Yet, the Korean state was still hesistant to divest from POSCO as demanded by the IMF since doing so, would heavily undermine the prospects of Korea's steel industry to mature. In any case, the IMF's real intention is not to help the Korean economy out of the doldrum. But allowing POSCO to be state-owned will not help matters either. Herein lies part of the limitations of the authoritarian Korean state's industrialisation policy. Having explained how the state in both Nigeria and South Korea constitute an obstacle to the industrialisation inclusive of steel development in these countries, it becomes a lot easier to understand other problems that they are faced with. It is to these other problems that we now turn.

ii] Conception of the Steel Companies: Within the two decades after attaining political independence, the structure of the economies of Nigeria and South Korea were not only basically agrarian and non-industrialised, but also, had a very weak productive base. Worst still, the local business class was not only biased towards exchange, but lacked the capital to embark upon industrialisation. Partly in an attempt to reverse the trend of economic underedevelopment inherited from colonialism in both countries, the state had to intervene into the economy. In theory, though the state had stated its commitment to the economic development of the post-colonial economies of the two countries, it was problematic in practise. The problem was not really that of failure on the part of the state as its path to economic development and conception of industrialisation inclusive of steel. Let us expantiate on this with its conception of steel development beginning with Nigeria.

In Nigeria, the public steel companies were established by the government in the hope that they could be eventually leased out to subsidiaries of steel transnationals and local private capital to manage and for which they would pay rent. Two major reasons were previously advanced why steel development was conceived in that context. First, industrialisation generally, and steel development in particular, were not really on the agenda of those in charge of the state. Second, as a political *alibi*, the managers of the state saw the steel sector as one of the newest project areas in the national economy which could be used to appriopriate wealth. This was so because industrialisation inclusive of steel development, still remained a major catch phrase that the political class would use to solicit political support from the people, and make it appear to them as if they were genuinely concerned with the improvement of their material well-being. Suffice it note that, industrialisation was not the primary reasons why the state embarked on steel development in Nigeria.

Viewed from this context, what observers and scholars of the Nigerian steel industry regard as problems to the steel companies like financial squeeze and lack of raw materials, are indeed, really no problem to the state. For, that is how the state wanted them to be: ridden with crisis to justify the allocation of more money in every financial year for their parochial gains. In fact, the crisis the public steel companies are faced with, largely mirrors the nature of state capitalism. This is one major reason why the previous efforts of the Nigerian government to privatize the steel companies had remained long-drawn and ridden with contradictions. For instance, the Nigerian government turned down the Hatch Report of 1988 in which it was recommended to the World Bank/IMF that for Nigeria to qualify and get their loan during the period of adjustment, the public steel companies should be sold to private capitals. It did so because it would have undermined the accumulative base of those in charge of the state and its cronies. In any case, the Bank and Fund were not sincere with their intentions as they wanted to corner the public steel companies for foreign steel capitals. Instead, the proposal of the state was to commercialise the public steel companies by leasing them to foreign companies for a period, while still retained full control of the companies. The state philosophy of commercialisation was contradictory. Because of this, Voest Alpine, an Austrian consortium that had indicated interest on DSC, backed out of the deal. The same was also true of Churchgate, an Asian conglomerate in Nigeria that had initially cashed in on its strong links with Nigeria's late junta leader, General Sani Abacha, and wanted to buy over DSC but stopped the deal after coming to terms with the underpinning political motives of the state. As for Ajaokuta Steel Project, the TPE of Russia, the major contractor company handling the uncompleted blast furnance, had indicated interest in completing the company's blast furnace on the condition that it would manage its operation for ten years. The government turned down the offer. No local and foreign capitals had shown real interest on the three inland steel rolling mills beyond the temporary measure of allowing some of the major distributors of these companies to bring in billets for rolling and for which fees were charged. While the government sought to guard its interests jealously, it shut down all the public steel companies indefinitely since 1996, arguing that it would wait to reactivate them when the economy recovers. After uncovering a scam involving the late General Abacha's family and the Ministers of Finance and Power and Steel in which they bought a $500million debt owed the Russian contractor company constructing the Ajaokuta's blast furnance for $2.5 billion, the regime of General Abubakar did not shown any interest in regenerating the steel companies. It is expected that the General Obasanjo-led government on taking over power from that of General Abubakar on May 29, 1999, will revisit the steel sector. This is more so since his junta in 1979, signed all the contracts of the public steel companies. Since the root cause of the crisis in the steel sector resides with the nature of the state and its mode of surplus extraction and not really that of a government, there is very little the in-coming government can do to revamp the ailing steel industry unless the Nigerian state is re-designed in such a way that its policies will be developmment and people-oreinted.[1]

Since the government was and still not interested in the private steel companies, it hardly made any attempt to help resolve the problems they were faced with. This is all the more so since the majority of them are foreign private steel companies and operated in guided secrets, making it difficult for government agencies, the Manufacturers Association of Nigeria, MAN, and researchers to have any information on them. For instance, although the major private steel companies like Universal Steel Company, Continental Iron and Steel Company both located in Ikeja, Lagos were members of the MAN, the Association did not have any data on them. And since the Nigerian state adjusted in June 1986, most of the private steel companies had closed their shops; yet, both the government and MAN were not aware of all this. Under this situation, it would be difficult for government to include them in its plans should it decide to formulate any steel policy.

In South Korea, conception of the steel companies by the state was different from Nigeria's. Rooted in Korean state's deepening industrialisation policy, the steel sector was planned to launch Korea's industrial and technological age. All this made steel development a 'national duty' which all Koreans were made committed to. Perhaps, this partly explained the dominance of the steel industry by the Korean government and the seriousness that was attached to steel development. Unfortunately, it is in the state's dominance of the steel sector that a larger part of the problems it is to be faced later, are to be located. In fact, the problem is not so much with the conception of steel development as a national duty; rather, it is the state mode of accumulation in the economy. Like the Nigerian state, its Korean counterpart is all pervasive and powerful. Once, it commenced steel development, it not only formulated policies to back it up, but also, almost monopolised the attraction of foreign credits for industrialization among others. In part, this made POSCO a steel giant in Korea, which was to the detriment of the private steel companies. With the imbalance in the Korean steel industry, POSCO alone, would be unable to lead it into market maturity in 2010 as planned. Not only that. The Korean government would have to divest largely in POSCO for it to be a private company and operate in line with the forces the market. In fact, this problem became evident since the IMF introduced conditionalites for bailing out Korea's traumatised economy. For the first time, the Korean government agreed in principle to restructure the equity of POSCO by inviting both local and foreign private capitals to acquire shares in the company. This, no doubt, will introduce a new dimension into Korea's steel industry if the state will really roll back from the mainstream of the economy.

Members of the Korea Steel Association, KOSA, who the author spoke to in Seoul in 1997, strongly condemned the steel policy of the Korean government for making the private steel companies almost wholly dependent on POSCO for most of their production inputs. They also accused POSCO which enjoys all government support for being political in most of its actions such as using its own production cost and prices for locally made steel products as a standard for the member-companies of KOSA, instead of allowing the market forces to prevail. To KOSA, all this has constrained the activities of the member-companies, particularly in terms of capacity utilisation and extent of their contribution to the industrialisation of the Republic of Korea. According to KOSA, one of the reasons for the protracted financial squeeze and eventually the bankruptcy of Hanbo Steel Company in March 1997, was because it intervened into the upstream sector of the Korean steel industry contrary to the government steel policy which made it exclusive for POSCO. And as a consequence, Hanbo Steel Company did not receive any credit from the government. This partly accounted for why the company's chief executive used the son of then President Kim Young Sam to secure a loan from Korean banks and subsequent scandal that followed it.[2]

iii] Finance: In Nigeria, the issue at stake is not so much with the protracted financial squeeze which the state-owned iron and steel companies have been faced with right from their inception. Rather, it is the root cause of the financial crisis, which is basically political. It is political in the sense that it originates from the nature of the struggle to accumulate between and among the various institutional groups that make the Nigerian state a reality and all of which helped frame the state's own conception of the steel companies. Since the public steel companies were ultimately conceived to generate rent through leasing, funding them after their completion was merely seen by the Nigerian government as underwriting their bills, but not really to invest. This made it pretty difficult for government to inject enough funds into the companies, and under such funding system, it was difficult for the management of the companies to plan effectively. Because the steel companies were not high on the agenda of the government, how much subvention they got from the government was, in most cases, tied more to its mood than the overall income accruing from oil export and production. Not only was the subvention to the steel companies irregular, the government even reserved the right to slash it down and also postpone its release indefinitely thereby crippling their operations. Some explanation is in order here.

Since the public steel companies were seen as rent-seeking projects its funding was characterized by large scale fraud as each of the state's institutional groups and its cronies struggled on grimly to get their own share of the shrinking national wealth most of which came from oil export and production. Little wonder then Minister of Steel under the Shagari-led government, Mallam Ali Makele of blessed memory had reportedly said that, 'steel must flow no matter the cost'. This was quite evident in the huge cost of constructing the public steel companies when compared with the cost of building similar size and capacity of steel companies elsewhere in Europe and Asia. According to the Minister of Power and Steel under the General Abubakar junta, Mr. Sulaiman, the Nigerian government had spent US$10 billion in the development of the public steel sector between 1979 and 1998.[3] Yet, a comparism of the cost of building the public steel companies with that of steel companies of the same size and capacity, and about the same period elsewhere showed that, it cost three times more in Nigeria. For instance, the same group of foreign contractor companies built Nigeria's DSC and Saudi Arabia's integrated steel plant of the size, capacity and technology. Whereas it cost the Nigerian government about US$2.5billion to build DSC as of 1982, the Saudi government spent about US$800million to get its steel company on stream in the same period. Worst still, at the Ajaokuta steel project in which about US$5billion was sunk, was still uncompleted as of May 1999. For instance, the construction of the blast furnace began in 1979 with parts whose life span was, in most cases, did not exceed 15 years. Yet, 20 years after, the plant is still under construction, which means that, most its production machineries were already obsolete before completion. Compared with South Korea's Pohang Steel Works of larger capacity which took three years to complete, it shows the hollowness in the approach the Nigerian state took to steel development. Still on Ajaokuta. The recent debt buy back of US$500million at US$2.5billion at the Ajaokuta steel project involving the family of late Nigerian Head of State, General Abacha and the Ministers of Finance and Power and Steel, A. Ani and B. Dalhatu respectively, was part of the scandal that has plagued the steel companies from inception. Not only that. In an attempt to amass wealth illegally, support services such as rail line which ought to be undertaken by the national government as part of infrastructural development, was taken as part of the cost of steel development. Even then, the ore-rail line project that is planned to link Itakpe with Ajaokuta and DSC was an after thought by the government.

As a result of the unclear focus of the state on steel development coupled with its own design to use the steel sector to justify ambitious budget most of which went into the private bank accounts of the state managers, the public steel companies suffered overinvestment. Not surprisingly, it was difficult to ascertain the real net worth of the companies, though the Power and Steel Ministry had put the figure at US$10billion. All this explained in part why no local and foreign investor was interested in buying shares in these companies when the Babaginda's government pretended to be serious with the commercialisation between 1987 and 1991. For, the real intention of government was not to sell any shares of the state-owned steel companies to the public since they lubricate state capitalist mode of accumulation.

All this contrasted sharply with the experience of South Korea. In Korea, the state had also spent about US$7billion for steel development ranking as one of the largest sectoral investment by the Korean government. Of this amount, US$5billion was spent in building POSCO, which was half the cost for developing the public steel sector of Nigeria, where the steel companies were not only of smaller capacity, but ridden with crisis right from inception. In fact, POSCO's installed yearly capacity of 41 million metric tons of crude steel was 20 times more than Nigeria's. Yet, the Korean government, which apart from the initial reparation of US$300million paid by the Japanese government with which it started POSCO, relied heavily on external loan to finance the development of the country's public steel sector. This was unlike in Nigeria where the public steel companies were built from the proceeds from oil production and export. Perhaps, one of the reasons for the unseriousness of the state with steel development is because the funds were not loans. It is not to assume that, if the Nigerian government had got external loan to finance its steel sector, all would have been well given the state's own conception of industrialisation.

Let us still dwell more on the financial crisis rocking the public steel companies in Nigeria to bring into greater relief, its contrasts with South Korea's. Shortly before the indefinite closure of DSC in 1996, then General Abacha administration had concluded arrangement to scrap the entire company alledging lack of fund as the sole reason. If the government had implemented the plan, that would have resulted in folding up the company. Yet, the same junta leader wanted to buy over Ajaokuta steel project with the hope of leasing it out to foreign capital to manage for a rent. His death on June 8 1998, saved the state-owned steel companies among other public companies from being converted into one of his assets. Still on DSC. The company commenced operation in 1982 without working capital, but with only commissioning raw materials provided by the contractor companies. This could sustain production for only three months, thereafter, the government had to procure raw materials for the company. Yet, all that the Nigerian government did, was to import iron ore and other consumables that could keep production for another three months, and after that, declined to provide further funds on the ground that there was no money. All this partly accounted for the company's very poor performance. For instance, except in 1984 when the company produced at 25 percent of its installed capacity of crude steel, its average capacity utilization was five percent before it was shut down indefinitely in 1996. The three inland rolling mills were equally badly hit by financial crisis. With the exception of Katsina Steel Rolling Company that was supplied billets by Kobe Steel of Japan based on special arrangement, which partly explained its relatively better production history than the mills at Jos and Oshogbo, they were all beset with chronic financial problem which impaired their ability to import billets. As explained earlier, the advent of the World Bank/IMF-led adjustment worsened the financial crisis facing the steel sector. Even since the official termination of adjustment in 1994, they were still no signs of their recovery in sight as they remained shut.

One thing is quite clear from the above analysis. That is, the real cause of the current financial crisis plaguing the public steel companies is not so much that the state was broke as its unwilling to finance them. Which means that it is not that the public steel companies are really unviable as the federal government would want us to believe. Also, this partly explains government's self-contradictory approach to the commercialisation and privatisation of the companies. In any case, except for the rolling mills, DSC which is not only the completed integrated steel company, but still at take-off stage, ought not to be considered for commercialisation in the first place. For, it is after the social and political roles which informed government's establishment of integrated steel company like DSC have been accomplished that the state should invite private capital to acquire some shares in the company. This was the approach the Japanese took to state's divestment from its steel industry, which both the governments of Nigeria and South Korea were still unwiling to adopt at the time this study was undertaken.

As for the Korean state, it had stayed too long in the country's steel industry that its divestment from the public steel sector had become pretty difficult. That is not all. Government's near monopoly of the Korean steel industry has obscured the real financial status of the state-owned steel companies. With particular reference to POSCO, while it is true that the Korean government had kept the company solvent since its inception, it has created the erroneous impression that it was about one of the most cost-effective steel companies in the world as most of its funds came from the government more so as it is state-owned, but not from the international stock market. In most cases, the funds attracted little or no interest since they were either sourced directly from the government, or came through it. All this makes it misleading to compare POSCO's present financial outlook with other steel transantionals like Nippon Steel of Japan. Rather, it is only after the Korean state has reduced its equity in POSCO to a minority with the public and private interests holding a majority that the company's market competitiveness and of course, its real financial strength will be understood. In fact, the IMF has, as one of its conditionalities for a rescue loan to the Republic of Korea, a considerable state divestment from POSCO and other public companies.

With respect to the private steel companies in Nigeria, they were also faced with severe financial crisis. Given the small equity base of these companies, the majority of them could not raise funds from the local and foreign stock markets particularly as they were not quoted in the country's stock exchange nor in any foreign stock market. Initially, some of them had relied on short-term credits from the local banks to solve most of their financial problems. However, with the advent of adjustment during which the value of the Naira [=N=], the official currency of Nigeria, was dramatically depreciated vis-a-vis other foreign currencies, banks were no longer lending to them and other manufacturing companies. In fact, the glut in the local steel market made it all the more so for the banks not to lend. During the period of adjustment for instance, only 5 out of about 30 private steel companies in Nigeria remained in skeletal operation as the rest 25 companies were forced to close shop either temporarily or permanently. Even after the official termination of the adjustment programme in 1994 by the junta regime of then Abacha, and since the inception of the Abubakar admninistration on June 8, 1998 following the death of Sani Abacha, the situation still did not improve as about =N=100.00 exchanged for US$1.00 as of May 1998. All has caused a dramatic increase in both the cost of importing steel raw materials notably billets, and finished steel products like steel bars, rods and flats. For instance, since 1986, CISCO, Ikeja, Lagos, has closed and re-opened its shops for more than five times, the major reason being lack of finance. As for Universal Steel Company, it first resorted to the recycling of scrap metals and stopped as the price per ton shut up from =N=1,500.00 before adjustment to =N=5000.00 during the adjustment period. Since 1994, it has been rising steadily, heating =N=7000.00 per tonnes as of 1997, thereby forcing the company to shut its production line of its mini-mill.[4] At the time this study was conducted, it was only barely importing semi-finished steel products like light plates and other flat steel of greater thickness for making enamel wares. Though both the public and private steel companies in South Korea had their own perculiar problems, they were different from their counterparts in Nigeria.

As noted earlier, POSCO's was seen not to be really involved in any financial crisis relative to its counterpart in Nigeria. This is largely because of the state's own conception of the steel company: investment in steel development. According to officials of POSCO, the Korean government had spent about US$7billion in the company and steel development generally, which represented its single largest sectoral investment as of 1996. That amount of public investment in steel represented about 3 percent of the Republic's GNP. POSCO's net worth was estimated at US$16billion as of 1996. Inspite of being a government company, POSCO was planned to declare profit, which was contrary to the view of the Nigerian government that its public steel companies should not make profits. So, in 1973, POSCO posted a net profit after tax of US$12million, and had been operating profitably since then. In 1978, for instance, it declared a net profit after tax of US$45 million and by 1995, it rose to US$1,050 million. One has followed convention so far. Some critical analysis is ideal here, especially to throw more light on my argument on the financial outlook of POSCO.

There is no doubt that the above financial figures presented POSCO as a buoyant company. That is by no means an assumption that there are no financial problems in the company, though they may not be comparable with the situation in Nigeria's public steel companies. As noted earlier, the Korean government had charged POSCO with the role of empowering Korea to compete with Japan's foremost steel giant, Nippon Steel Company, NSC. It was also planned to catalyse Korea's industrialisation so that it could as well, compete with Japan industrially. These are all mistaken positions. For one thing, while POSCO was still a state-owned company, the NSC had long been a steel transnational with the majority of its equity shares largely held by private investors. The same is true of other steel transnationals like Unisor-Sacilor of France, Riva of Italy, Krupp of Germany, and Arbed Group of Luxembourg with which POSCO has been mistakenly compared. With about 95 percent of the shares of POSCO still held by the Korean government and its agencies like the Korean Development Bank, it will be out of place comparing it with the NSC of Japan as the Korean government and some scholars of the Korean economy have always done, since no matter how financially buoyant POSCO might be, it is largely a public company enjoying all the privileges of the state like cheap funds and protection. It is still in doubt whether if any of the government loans to POSCO was ever repaid. The politics of the financing of POSCO has obscured its correct financial position and market competitveness. In essence, it is only after the Korean government had reduced its shares in POSCO to a minority position and begin to allow its operations to be guided by the market forces more so as it is already in the stage of late growth, that its real financial strengths and weaknesses would emerge. The fact is that the Kim Dae-Jung government was still reluctant to divest substantially from POSCO in line with the IMF's bail-out conditionalities as of the end of 1998. This has reinforced the fears already being expressed in some official quarters in Seoul that, should the government divest significantly from POSCO as the IMF had dictated, that would bring it back to its real level inclusive of its financial position, a situation that would upturn the capacity of the Korean steel companies to compete in the global steel market. However, leaving POSCO still dominated by the government would not really help matters particularly if the country's steel industry is to attain market maturity. Herein lies the limitations and politics of the current financial capacity of POSCO. Indeed, if the Korean economy did not trough in late 1997, definitely this is one major problem that would have still constrained the Korean government effort of moving its steel industry into market maturity by 2010 as planned as the state and the market are incompatible.

In Korea, the private steel companies were, like their counterparts in Nigeria, faced with financial problems. As noted earlier, Korea's policy of 'deepening industrialisation' in the late 1970s, compelled it to invest heavily in steel development. But this was done to the detriment of the private steel companies, which, unlike the big business groups, did not benefit from the cheap credits that the Korean government advanced to the heavy industries in the 1970s and 1980s. In other words, the private steel companies had had to contend with the problem of raising their investment capital right from the onset. And since virtually all banks in South Korea were either owned or controlled by the government, it was difficult for the private steel companies to source for loans. Where it was possible, the lending rate was so high that they abandoned it. This was perhaps one of the reasons why Hanbo Steel Company had to give a gratis of US$200million to the son of then President Kim Young Sam in order to get a bank loan to finance the company's acquisition of production machineries for the new steel plants and other expansion projects.

Though POSCO was expected to meet most of the basic raw material needs of the private steel companies, they still had need to import some of them like the initial production machineries. According to KOSA, its member-companies had no option but to import the main production machineries that POSCO either could not produce when they took off, or still did not produce them long after coming on stream. Yet, the Korean government failed to extend its relatively cheap loans to the private steel companies to help boost the local production of intermediate and capital goods. As a consequence, most of the companies could produce only 30 percent of their installed capacity, with little or no hope of increasing their output since the onset of the financial crisis rocking Korea's economy since December 1997. That is not all, unlike POSCO, the private steel companies hardly had political links that they could use to facilitate bank loans especially as virtually all banks in Korea were owned by the state. As noted, it was only Hanbo Steel Company out of the over 100 member-companies of KOSA, however, that succeeded in securing a foreign loan to set up its integrated plant without government backing. The scandal and sunsequent bankruptcy of the company that followed the loan were already well publicised in both local Korean media and internationally to merit further analysis here.

One fundamental issue that has emerged so far on our analysis of the financial crises facing the steel industries of both Nigeria and South Korea is that, there is a limit to which the state can indeed, invest in steel beyond which it becomes counterproductive. The state should start off the steel companies, nuture them to the stage of late growth, thereafter, withdraw to playing a regulatory role. It is expected that at the late growth, the state should have used the steel sector to accomplish most of its social projects such as the construction of social infrastructures like roads and public buildings among others. Thereafter, the public steel companies could be privatised and it is expected that other private companies should have been fully developed. Together, they would take the steel industry into market maturity. More details on this in the next chapter.

iv] Management Crisis: To a large extent, the management techniques adopted by the public steel companies in both Nigeria and South Korea were ridden with contradictions, contradictions that stemmed from the state's own conception of these companies and mode of surplus extraction from the economy. In most cases, it is these contradictions that really present themselves in the form of management crisis, among others. Which means that attempts at resolving them should begin with the tackling of the root cause[s] of the crisis. Let us shed more light on this by examining the management crisis in the steel companies of both Nigeria and South Korea, beginning with the former.

In Nigeria, the public steel companies were set up as parastatals, yet they pretended to be managed commercially. As parastatals, the thinking of government is that they should not be concerned with declaring profits, but to create jobs, promote even economic development, among other social functions. Because they were funded by the government also, more of political than economic calculations were ultimately factored into their management.

Yet, the government would want the public steel companies to operate like joint stock companies and declaring profits. Hence, soon after they were commissioned, the government reminded the management of the steel companies of the need and urgency for them to be self accounting, instead of waiting for state subvention. But any attempt by any of the executive of these companies to syndicate loans from the bank in the face of creeping financial crisis, was aborted by the same government. For instance, the attempt by the management of Oshogbo Steel Company to secure loan from a Nigerian private bank in 1987 with which it wanted to import billets was thwarted by the federal government. That is not all. The five public steel companies were established by government in 1979 with an equity of =N=2.00 each, bringing the total to =N=10.00. This was an equivalent of about US$5.00 as of 1979. But in actual fact, about US$10billion or more was already spent on them as of 1998. This was just an aspect of the marriage of two strange management styles, which has brought severe management crisis in the public steel companies.

The contradictions arising from the management crisis are many and grave. Although the chief executives of the public steel companies were mandated to oversee their day-to-day operations in consultation with their board of directors, they were in most cases, controlled by the superivising Ministry of Power and Steel and other political heavy weight both in and outside of the government. Quite often, chief executives of these companies were told at one time not to concern themselves with declaring profits and at another time, to operate commercially, meaning making profits. As chief executives of these steel companies, the government still dictated to them where to procure their basic raw materials from and at what price even when experiences in the past had shown that, the price of raw materials sourced by government was higher than that obtainable in the open market; or lower still, if the companies were given the freedom to negotiate the price elsewhere. This was more glaring when the coke oven battery of the Ajaokuta steel project was commissioned in 1991 by the Babangida's regime. Then military admninistrator of Ajaokuta steel project, who was an officer of the Nigerian Airforce, was handed a list of government appointed suppliers of coke, a list that was dominated by the fronts of top military officers and the oligarchies. It was yet another scam in the steel industry which was similar to the cement armada of the 1970s. About 1.5billion naira was spent on importing coke to ignite a coke oven that was built with 120million naira. Not only that, while a majority of the contractors who were fully paid for the supply did not deliver, the few who did, supplied already caked coke in addition to the difficulty of evacuating the coke to Ajaokuta forcing them to resort to the use of trailers as opposed to rail or sea. In all this, the chief executive was more of transmission belt to the powers that be, though he could not have been left without being settled. Government also fixed prices of the rolled products arbitrarily even when such prices which in most cases, were either far below or below the cost of production, depending largely on what they stand to gain from the price changes. For instance, one tonne of rod from DSC and the three steel rolling companies was sold at 16,000 naira in 1986 compared with 13,000 naira per ton of the same qunatity of rolled product. This made the public steel companies to suffer a glut since steel consumers preferred to buy from the steel merchants. With the inability of the public steel companies to sell their products, the custodians of the state would in addition to the allocations made to themselves, now share the supposed unsold rolled products among themselves, which, they, in turn, sell to the local steel mercahnts at lower a price. Most often, it was the allocation papers that were sold to the steel vendor who now collect the products.

What is more. All senior appointments made by the companies would have to be confirmed by the government as means of protecting the parochial interests of the state. In essence, major appointments in the steel companies are not only political with its concomitant effect on the efficiency as merits is usually undermined. One of the consequences of prolonged excessive government interference into the internal affairs of the steel companies, is that they still had no discernible structure for managing them at the time this study was conducted. This, in turn, had affected the performance of the steel companies as their chief executives were not allowed to put in place a structure that would enable them to operate commercially or as parastatals.

With respect to the private steel companies, the real problem is not so much with the management as their nature of ownership. As noted earlier, the real shareholders of the majority of the private steel companies were still kept as top secrets. It was not perculiar to the steel sector; for, this was generally characteristic of most businesses in Nigeria like textile industry, in which Asian capitals were involved and dominated as well. Anyway, all this was done with the connivance of some Nigerians who acted as fronts. In fact, the author had garthered from some indigenous middle management staff of Asian-owned steel companies operating in Nigeria like CISCO, and USC that, the majority of these companies were really no less than a family business and of the single proprietorship type, though they operated as if they were limited liability companies in Nigeria. Therefore, one of the identifiable problems emanating from their management approach resides with the confusion arising from running a one-man business like a joint-stock company. This, itself, had made it difficult to have a fair idea of the overall performance of the private steel sector in particular and that of the country's entire steel industry generally.

In South Korea, the management style adopted for the running of the public steel companies was almost directly the opposite of Nigeria's. Korea's public steel companies, notably POSCO and its subsidiaries, were all run as limited liability companies whereas they were no less than public enterprises with the state fully in control of them. Though they were state-owned steel companies, the state did not really interfer so much into their day-to-day activities as it was the case in Nigeria. Without caring much about the structure of ownership of the public steel companies in Korea, they operated as if they were steel transnationals. As of 1995 for instance, POSCO's cost of producing one metric ton of crude steel was US$450.00. Compared with the cost per metric ton of crude steel of US$620.00 in Japan, US$520.00 in the US, and US$510 in Taiwan, POSCO was ajudged the most efficient steel company in the world in 1995. That was quite misleading since Korea's steel industry would still take two decades or more to arrive at market maturity, the stage where Japan's steel industry has been for about twenty-five years.

No doubt, POSCO's impressive performance over the years has led many to misconstrue its management technique as one of the most efficient in the world, whereas it is not as its operations were still guided by the forces of state capitalism and not those of the market. As the author argued earlier, it is after the company has been fully privatised, and not in its current status, that Korea public steel companies' management style could be set free from a blend of state-owned and joint-stock company approach. For, as officials of POSCO who the author spoke to in Seoul in 1997, had put it, 'the state still decides who sits on the company's board of directors, appoints its chief executive, and approves its policies and projects before they are embarked upon'. This is one of the problems that the government of Kim Dae-Jung has been grappling with since he started the partial implemention of the conditionalities of the IMF's financial rescue in 1998.

Unlike in Nigeria where its private steel companies were largely of the single proprietorship type, in Korea, they were larger in size and therefore, managed as joint-stock companies. That the private steel companies in Korea were more developed and larger in size, should not be construed delete that the family factor was not there. It is a major problem across the industrial sector of the Korean economy, including even most the *chaebols*. According to KOSA, there was virtually no private steel company in Korea whose origin could not be traced to one family and still largely managed as such. As a part and parcel of a larger family business for instance, the majority of the private steel companies were run not really as limited liability companies, nor with a stinct of efficiency as some of the workers, no matter how unproductive, were hardly fired. Nor was the operation of the companies really guided by modern management techniques. On the contrary, they were run, in most cases, on family sentiments. Hence, there is a considerable measure of loyalty and commitment from the employees to the companies since most of them were familially linked together. This had made it difficult to really run the majority of the private steel companies as shown in Chapter Three as purely limited liability ventures, though their books pretend to show that they are managed as such. They hardly adopt modern management techniques like management by objective. Again, the author had observed that due to family ties, workers hardly saw the essence of having insurance policies, nor did they have a retiring age. All this shows that, a lot is still left to be desired from the way Korea's private steel companies among other sub-sectors of the industry, were managed.

v] Crisis of Intersectoral Linkages: This was more of a problem with the Nigerian steel sector than Korea's. And the basic reason why it was a problem in Nigeria was because steel development was not really on the agenda of those in charge of the state. Hence, the steel companies were built without an industrial policy acting as a guide. This contrasted with the approach which the Korean government took to steel development. Let us explain how the poor interlinks between and among the steel companies, and with the rest of the economy constituted a problem to steel development in particular, and Nigeria's quest for industrialisation generally.

The problem of poor backward integration was both within and between the companies that make up the steel industry; and between them and the rest of the country's economy. Between and among the steel companies that constitute the steel industry, there was hardly any linkage as of the time of writing. Given the abundance of steel raw materials that abound in the country, the integrated steel company, notably, DSC ought not to suffer from acute shortage of iron ore, coal and alloying minerals. In other words, if DSC was still faced with chronic shortage of raw materials like iron ore long after it came on stream, and NIOMCO, Itakpe too, had begun production, then it can be said that there is poor linkage within the steel companies. For instance, the government began the importation of iron ore for use at DSC in 1983 after the commissioning quantity of iron ore brought in by the contractor companies was exhausted. In 1983, 215,000 metric tonnes were imported; 353,000 metric tonnes in 1984; and in 1985, 393,000 metric tonnes. When the state adjusted in 1986, the quantity fell to 202,015 metric tonnes and by 1990, it was 195,225 metric tonnes. Even these quantities could only sustain production for a maximum of 21 days of double shift. The trend of decline in imported iron ore continued into the rest of the 1990s with no importation at all in some years, before the company was shut down indefinitely in 1996. Yet, the government could not channel the beneficiated iron ore from the skeletal production from Itakpe to DSC. In fact, it was only after the beneficiation plant at Itakpe was installed that the government began the construction of a rail line from the Itakpe to Ajaokuta and DSC. That is not all. Without an industrial policy and coupled with poor planning, the government did not consider it wise to first commence the exploitation of high-ferrous grade of the huge iron ore deposits at Chokochoko and Agbaja both of which are about 70 kilometers away from DSC. DSC took off with 100 percent dependent on imported iron ore not because the state is unaware of its implications for the country's quest for steel development; rather, it profited its mangers to do so. The crisis has also plagued the exploitation of coal, another major raw material for iron and steel making.

That no consideration was given to the exploitation of the Obi/Lafia coal deposits by the government does not mean that the local huge reserve of coal that abound in the country was unsuitable for iron and steel making. Rather, the state managers were more interested in the temporary self enrichment from the importation of coal than the long-term benefits from developing the local reserve. This was clearly demonstrated in the scramble for contracts to import coke for the commissioning of the coke oven battery of Ajaokuta steel project, which was lit since 1991 while the blast furnace it was expected to feed, was still uncompleted 20 years after its construction began in 1979. Findings from the testing of the Obi/Lafia coal at the pilot coke oven of NMDC, Jos, showed that, if it is blended with small quantity of coal imported from Australia, the matallurgical value of the coal would be enhanced to a level that it is suitable for iron and steel making instead of abandoning it and resort to large-scale importation of coal. In fact, this is how politics of state accumulation has caused acute shortage of steel raw material crisis in Nigeria's steel industry.

In terms of complementarity, the public steel companies and other government agencies in Nigeria's iron and steel sector were hardly interlinked. If for anything, they duplicated their functions. The problem is not really transforming units of then National Steel Council into sub-ministries, rather, it is the politics of it all, which has turned out to be a constraint to steel development. For instance, what came to be known later as NIOMCO, Itakpe, was originally, the mining unit of then National Steel Council, Kaduna. It, however, only became autonomous and independent of its parent body, NSC, for no other reason than the plot by the state managers especially the bureaucratic class to turn such agency into another rent-seeking oufit. The same was true of SRMEA, Kaduna; and NMDC, Jos. The arbitrary creation of these agencies by government finally turned out to be counter-productive. For instance, neither DSC nor Ajaokuta steel project ever worked in close concert with any of these government agencies even when they ought to complement each other if all had gone well with steel development in the country. The SRMEA, the exploration unit of then NSC, hardly sent samples of its mine to NMDC for laboratory analysis. Rather, it has a mini laboratory where its samples were analysed. NMDC has, on its own, got samples from the field for analysis. The findings from the NMDC's study was still left in the file as of May 1999. NMDC conducted a pilot research on the coakability of the Obi/Lafia, and its finding was positive. Under this situation, it will be difficult for the public steel companies to ever perform.

If between and the public steel companies, there was no links, it is natural therefore, to expect them not to be linked with the private steel companies. As explained earlier, the state did not include the private steel companies in its calculus when it established the public steel companies. Not surprisingly, the private steel companies were left to fend for themselves both in terms of sourcing for their basic raw materials and other production inputs, and funding. For instance, there was hardly any form of collaboration between the government-owned steel agencies and the major private steel companies like USC and CISCO, both located in Ikeja, Lagos. And as a consequence, the private steel companies had to import most of their basic raw materials with the remainder sourced through recycled metal scraps. They also imported finished steel products like rods, coils and plates to produce intermediate and capital goods. Though operating within one economy, both the public and private steel companies were independent of each other, and at times, seek to undermine each other's capacity to survive. For instance, during the period of adjustment, the private steel companies imported finished steel products like rods and coils which were sold at 12,000 naira per tonne compared with 18,000 naira for the same quantity of product. At the price the private steel companies sold their products, it was difficult for the public steel to market their products thereby creating an a glut in the local steel market. The private steel companies were able to undermine their counterparts in the public steel sector due largely to the absence of a steel policy. The fact that they have the backing of some influencial government oficials made it all the more so. All this contrasted sharply with Korea's experience with iron and steel development.

In South Korea, steel development, as pointed out earlier, proceeded under a regime of an industrial policy that made the country's steel sector as its bedrock of industrialisation. That is, to say that, Korea's steel industry helped restructure the path which the government later took to industrialisation since 1961. One of the traditional functions of the steel sector of a country is to serve as the bedrock of its industrialisation project. Which means it is expected to catalyse its industrialiastion by producing the basic raw materials either in finished, or semi-finished form for the intermediate and capital goods-producing companies. Also, most of the products of the integrated steel companies are themselves, intermediate and capital goods. In terms of its interlinks with other manufacturing companies, POSCO's product mix was specifically planned in such a way that the downstream companies should be able to source for raw materials from its wide range of steel products. For instance, POSCO produced flat steel on specification for the Hyundae Shipbuilding Company. Still on specification, POSCO produced flat steel of various thickness, ductility and length for the majority of Korea's automobile-producing companies. The majority of Korea's fabrication engineering companies were equally largely dependent on POSCO for their basic production inputs. In fact, with the exception of the microchips-producing companies in South Korea, there was virtually no intermediate and capital goods producers that did not significantly rely on POSCO for most of their raw materials. This is the nature of the interlinks between the steel companies and other sectors of the Korean economy, unlike in Nigeria where they operated as if they were an enclave unto themselves.

One interesting aspect of the backward linkage of Korea's steel sector with the rest of the economy is the government's policy to substitute incrementally, hitherto imported intermediate and capital goods, with the locally produced ones over a given period of time. For instance, since the implementation of policy of deepening industrialisation in the late 1970s, a target of 100 percent reduction in imported steel products like rods and coils over a period of 25 years. By 1993, South Korea had achieved a self-sufficiency of 100 percent in steel rods and coils; and about 60 percent self-sufficiency in flat steel products. The impressive performance of the Korean steel industry in this respect, was due largely to wide range of products rolled out of POSCO. All this made it possible for the establishment of downstream private steel companies that use products of POSCO to manufacture parts for the Korean railway industry, auto-giants and heavy machineries while creating jobs in the process. This portends great policy lesson for Nigeria.

With an industrial policy that placed high premium on self-sustainable industrial and economic growth, and a committed post-colonial state, South Korea was able to significantly turn around, the lopsidedness that charaterized the industrial sectors and the larger economy of South Korea for 54 years of self rule. However, the capacity of the Korean government to actually sustain the use of its steel sector to link up with the rest aspects of its economy largely depends on how long the state will want to remain both as an active actor and regulator in the real sector of the Korean economy. Given the wave of globalisation across the world and the relatively fragile economy of Korea in comparison with some of the major industrial and market economies like Germany and Japan, there is every tendency for the state to want to remain at the heart of the economy inclusive of steel. In which case, its steel sector that has already attained late growth, might begin to stagnate. In fact, the current structural crisis that the Korean economy has been faced with, and the half-hearted efforts made by the Kim Dae-Jung government to deal with its root cause by partly implementing the conditionalities of IMF, is an indication that a lot more needs to be done to restore it back to its right footing.

6.3 Prospects

Inspite of the problems the steel industries of both Nigeria and South Korea are faced with as explained above, there is still great prospects for their survival. Given the huge and untapped local market for the local production of intermediate and capital goods in both countries, there is a bright future for their steel industries provided the right economic policies and mixes are in place, and with a conducive political atmosphere.

In Nigeria, particularly since the turn of the 1980s, there has been an emergence of indigenous and privately owned small equity-based intermediate and capital goods-producing companies most of which were located in the eastern part of the country such as Nnewi, in Anambra State. They specialised in the fabrication of spares for the auto-industries. The fact that these companies could spontaneously spring up to take advantage of the huge untapped local market for auto-parts in the country showed that, if the Nigerian government had had an industrial policy, they would have done better and even helped a great deal to quicken the process of Nigeria's quest for industrialisation. One thing the government should do in order to have these private intermediate and capital goods producers contribute more to the country's industrialisation, is to formulate a steel policy that compels the steel companies to produce most of their basic production inputs. If the integrated steel companies are made to roll out flat steel products among other intermediate and capital goods, this would go a long way in reducing the import dependency ratio of the companies. For, it is only when the steel companies have a wide range of product that they can promote intersectoral linkages within the steel industry, and between the steel industry and the rest of the economy.

Unlike South Korea which is not endowed with the basic steel raw materials like iron ore and coal, and alloying minerals, Nigeria's steel companies ought not to have suffered from acute shortage of these raw materials since they abound locally; and that if the state had had an industrial policy that guided its steel development project, all would have been well. Suffice it to note that, the nature of the politics of the Nigerian state and its mode of surplus extraction are the root causes of why steel development has not really got underway. Once the state managers are re-oriented away from using state power to actualize parochial gains, and ensure that national policies are people oriented, then Nigeria's quest for industrialization would begin to redress the problem of dependence.

This, in turn, means that the state would have to rethink its own conception of steel development in particular by formulating an appropriate steel policy that will enable the steel sector to play its role as the catalyst for the Nigeria's industrialiastion as obtains in some of the major industrial economies like Germany. Once this is settled, coupled with the exploitation of the vast untapped deposits of iron ore, coal and alloying minerals in the country, Nigeria should be a major steel producer both within Africa and the world at large.

Despite the fact that steel development has never really been on the agenda of the Nigerian state, the country's political system made it all the more so. Since its political independence in 1960, Nigeria has only been under civil rule for nine years, meaning that the military had governed the country for three decades. And from the past experience of the junta in Nigeria, industrialisation had hardly entered into its calculus of planning. Quite often, the problems in the economy which the military gave for its intervention into the political terrain were, rather than solved, compounded. Its lack of training in democratic governance worsened matters. And as a consequence, the country's industrial sector in particular and the economy at large were worse off than what they were before the military intervened into politics in 1966. For, far from being concerned with putting the country's steel and industrial sectors back on keel, the military regimes have, on balance, presided over their underdevelopment. Returning Nigeria to civilian regime by May 29 1999, does not mean that all will be well with steel development. Rather, the expectation is that, an atmosphere for dialogue on the major problems facing the Nigerian economy would have been created as opposed to the authoritarian approach of the military regimes. As noted earlier, it expected that the Nigerian state under a civilian dispensation, would be re-organised in such a way that its policies sholud be concerned with the enhancement of the well-being of the majority of Nigerians. As part of the re-organisation, the state should be made less attractive politically. Then and only then, would one really begin to think in concrete terms, of the prospects of Nigeria's steel industry.

The factors that shape the future of the Korean steel industry are different from Nigeria's in many repects. Nature has made the upstream sector of Korea's steel industry almost wholly dependent on import, unlike in Nigeria where the basic steel raw materials abound. Compared with Nigeria, one of the unique features of the Korean steel industry is that, inspite of the country's poverty in steel raw materials, the Korean government was able to get its steel industry underway and even used the sector to put its industrialisation policies and actions on a relatively better footing. In an attempt to ensure a steady supply of basic steel raw materials for POSCO, the Korean government had signed joint venture agreement with the Australian government for the joint development of iron ore deposits in Australia. The Korean government did the same with the Canadian government for the development of coal in Canada. These were the two major sources from which POSCO procures its major raw materials as already explained in Chapter Three. Which means that Korea's high import dependency ratio in terms of raw materials places the future of its steel industry at the mercy of the foreign iron ore and coal-producing countries. Should Australia and Canada decide to discontinue with the joint exploitation for instance, that would disrupt the activities of not only POSCO but other downstream companies that are dependent on the major Korean steel company. That they have not done so in the past does not rule its occurrence in the future. All this might limit the pace at the which Korea's steel industry ordinarily would have wanted to grow.

Apart from the natural limitation that Korea's steel industry is faced with, the private steel sector has been particularly relatively underdeveloped. This is more glaring when it is compared with the public sector of the industry. For instance, the Korean government's decision to discourage the private steel companies from intervening into the upstream sector in order to save them from facing the high import dependency ratio for steel raw materials is a tab on their potentials. In fact, this has been vehemently criticised by the member-companies of KOSA. The main contention of the Association to which the author partly subscribe, was that, Korea's private steel companies have, by the steel and industrial policies and actions of the state, had their operations and capacity to expand severely limited. The implications of all this, is that, the envisaged market maturity of Korea's steel industry in 2010, might remain elusive since the public steel sector alone can hardly lead the country's steel industry to such stage of growth.

6.4 Concluding Remarks

From foregoing analysis, the steel industries of both Nigeria and Korea were, in no doubt, operating at different stages of development. So, too, were their problems and prospects. At this stage, the pertinent question that arises is whether there is really anything for Nigeria in the Korean steel industry. This and other related issues are addressed in the concluding chapter.

Notes and References

1. What the General Abubakar's administration was able to do in terms of regenerating the ailing public steel companies and agencies between June 1998 and May 1999, was to inaugurate in March, a National Inter-Ministerial Committee on the Completion of Iron and Steel Projects. The Committee was charged with the duty of recommending measures to the government on how to complete the outstanding projects and rehabilitate the entire industry. The author was invited by the Minister of Power and Steel to brief the Committee on the experience of South Korea with steel development. From all appearances, the junta only engaged in defensive radicalism by pretending to be serious with steel development but it is not.
2. Based on the author's interviews with the officials of KOSA in May 1997.
3. The real amount spent on the public steel companies and agen-

cies by the government has been mired by controversy. Compared with steel companies of the same size and handled by the same contractor companies, investigations have shown that the state-owned steel companies cost three times more. For instance, out of the US$10billion government claimed to have spent on the steel companies, agencies and support services so far, not more than US$6billion was spent; the remainder was believed to have gone into the private purse of the state managers and their cronies. The contracts for the steel projects were signed by the Murtala/Obasanjo junta in 1979 when one naira exchanged for US$1.5.

4. Based on the author's field trips.

Chapter Seven

Conclusion: Policy Recommendations

7.1 INTRODUCTION

In this chapter, one will briely revisit some of the theoretical issues that underpin industrialisation including steel development in both Nigeria and South Korea, then proceed to make policy prescriptions. Before the discussion proper, it is pertinent to first present an overview of the current financial crisis which South Korea has been faced with since late 1997, noting its implications for Nigeria's quest for industrialisation.

7.2 AN OVERVIEW OF SOUTH KOREA'S CURRENT ECONOMIC CRISIS AND ITS POLICY IMPLICATIONS FOR NIGERIA'S QUEST FOR INDUSTRIALISATION

How come that the impressive growth rate which South Korea's economy has been recording since the turn of the 1980s till the early part of the 1990s could not prevent the Republic from sliding into economic crisis in late 1997? This is one of the fundamental questions that continues to agitate the minds of scholars, policy makers, government officials and other keen watchers of the political economy of Korea. From the received literature in the field, opinions are polarized into two broad and opposing groups as to the origin and nature of the Korean current economic crisis as well as the probable ways out of the doldrum. To the first group of scholars like F. Deyo [1] and R. Kiely [2], most of whom have been ardent critics of the development strategies adopted by South Korea and other Asian Tigers namely, Taiwan, Singapore and Hong-Kong, Korea's recent economic crisis was already predicted since the 1980s given the inherent contradictions in the path that it took to economic development including industrialisation. In fact, Paul Krugman had, in one of the most recent writings on the Korean economy, shared the same pessimism with other scholars of the Asian NICs, arguing that there was never anything 'miraculous' about South Korea's growth. Suffice it to note that, contrary to the view being spread by the IMF and other apologists of the Korean development strategy that the crisis which the Korean economy is currently immersed is financial, Deyo, Johnson, Krugman et al contend that, it is indeed, economic and represents no less than a vindication of earlier fears expressed by these scholars. Fears that the Korean economy would sooner than later, collaspe if its mono-growth development strategy is continued with. The author partly subscribes to this position. For one thing, the origin of Korea's recent economic crisis is not only traced to the nature of the Korean state particularly since 1961, but also, has cast serious doubt on the World Bank's survey of 1993 in which the Bank had hailed the economic policies of the Asian NICs and recommended them for Africa's adjusting economies recovery strategies.

To the second group of scholars notably, M. Feldstein[3], D. Henderson,[4] D. Brautigam [5] and S. Rodelet et al,[6] they think otherwise, arguing that, Korea's recent crisis was not only financial in nature, but one of those temporary problems that accompany any fast growing economy. And that sooner than later, the Korean economy would pick up if the Kim Dae-Jung-led government would fully implement the IMF's bail-out conditionalites. Perhaps, this partly informed the shared optimism among these scholars that, with the cessation of the economic crisis in no distant future, the production system of the economies of the Asian NICs of which South Korea is about the most important member, has great propects of taking over the sustainance of the growth of global capitalism from Western Europe and the USA in the next millenium.[7]

As explained earlier in Chapters Two and Three, inspite of the US long strategic presence in the Republic of Korea, the Republic

Conclusion

is still far from getting imbued with American culture and politics. The Republic's production system can hardly be described as really capitalist in the Western sense of it. So, too, is its democracy. Korea's production system can at best, be better described as 'Confucian capitalism', and its political system 'Confucian democracy'. All this makes the optimism shared by these scholars misleading. One of the reasons for this, is the inability of this group of scholars to develop an alternative rigorous theoretical basis for analyzing the production system and the inherent contradictions of the Asian NICs.

Not only that, as in spite of the debates that have flourished on the Korea's economic crisis so far, and some of the efforts being made largely by the IMF to deal with it, there is little or no hope of real recovery in sight. Why? The root cause of the Korean economic crisis stems from the contradictions arising from the nature of the Korean state and its mode of surplus extraction. Which largely explains why Korea's crisis may present itself as economic, it is, indeed, political. And as a result, the Kim Dae-Jung administration is seeking for an expedient way out of the doldrum instead of dealing decisively with its root cause. This is because tackling the root cause of the problems in the Korean economy would require a major re-organisation of the Korean state and its model of capitalism along democratic lines which Kim's government has not the political will to embark upon. Since the Kim's government began the partial implementation of the IMF's bail-out conditionalites since the turn of 1998, there has been more down-sizing across all sectors of the economy, withdrawal of state's financial support to the majority of the *chaebols* resulting in their liquidation, and gradual attempts at privatisation of companies hitherto wholly owned by the state or those in which it controls majority shares. All this has reinforced the strong feelings by some government officials and scholars in Korea that, the IMF's rescue may, on balance, further deepen the crisis in the Korean economy in the long term. Apart from accusing the IMF of stepping its traditional function, there is strong suspicion among some Korean academics and policy makers that, the Fund only took advantage of the Korean crisis to force the Korean government to open up its relatively closed economy to exports from Japan and other member-countries of the Organization of Economic Cooperation and Development,

OECD.

Furthermore, if the Korean economy is still closed to the exports from the Western European countries and Japan long after joining the OECD, and also, having ratified the treaty establishing the World Trade Organization, WTO, then a lot is still left to be desired about the Korean government's industrial growth and preparedness for globalisation. In fact, there are strong feelings among some Korean government officials and scholars too, that the so-called globalisation project of then government of Kim Young-Sam, which is being continued with by the regime of Kim Dae-Jung, has remained only on papers since the government has not really put any globalisation economic project underway. This should not be surprising as Korea's economy generally and its major industries in particular, were still largely owned, managed and funded by the state in most cases. It is for this and other related reasons that the Korean government has been examining globalistion in the past five years with utmost suspicion in order to prevent foreign capital from taking over its economy. For instance, since the Kim's regime began the partial implementation of the IMF's conditionalities in 1998, foreign capital has, together with the Fund, been mounting pressure on it [the government] to privatise POSCO and the big trading companies. The reluctance of the Korean government to the full privatisation of POSCO was similar to Nigeria's when then regime of General Babangida embraced the World Bank/IMF-led adjustment programme in 1986. Like the Nigerian government, the government of Korea still wants to control POSCO largely because it helps lubricate its own model of accumulation; hence, it is not really interested in divesting significantly from the steel company.

From the above analysis, the major issue at stake is not really how the IMF would help bring about a temporary cessation of the economic crisis in Korea. Rather, it is the identification of the root cause of the crisis, and then, the mapping out of the right strategies plus the political will to deal decisively with it. What all this portends for Nigeria is that, inspite of the rapid industrial and economic growth that Korea had recorded in the past 25 years, the unfolding contradictions call for proper scrutiny of the Korea's development strategy. Let us now return to the discussion proper.

7.3 WHAT IS IN KOREA'S STEEL INDUSTRY FOR NIGERIA?

To be able to adequately address this question, it is important to return to the philosophical context within which the Korean state embarked on industrialisation inclusive of steel development. For, it is with a deeper understanding of the basic theoretical issues that underpin Korea's industrialisation that Nigerian policy makers, bureaucrats and scholars would be able to rethink Nigeria's quest for industrialization with the hope of coming up with articulate but practicable national industrial/steel policy and projects.

7.3.1 Philosophical Context

Development Strategy: Since one has already clarified basic unsettled theoretical issues that underline industrialisation in Nigeria and South Korea in Chapter Two, they will only be revisited to provide a background context to this section. In doing so, cognizance is taken of the strengths and weaknesses of the policy issues in the industrialization of both countries particularly as part of the lessons for Nigeria.

There is no doubt that South Korea had witnessed rapid economic growth compared with Nigeria, and even for the rest of Sub-Saharan Africa. The theoretical discourse on the distinction between growth and development is so well treated in the received literature in the field that it should not detain us. However, one vital fact to note from the theoretical discourse on the political economy of development of developing economies is that, it is possible for an economy to grow, just like as the Korean economy did, without really experiencing development. In fact, that is what the underdevelopment and dependency theorists' [UDT] major thesis of 'growth without development' is all about. That partly explains why it should not be wildly assumed as most apologists of the Asian NICs did, that the dramatic growth experienced by the Korean economy in the 1980s and up till the early part of the 1990s, was a celebration of the liberal economic theories. For, from a closer look and more critical analysis of the Korean production system including its approach to industrialization, economic development understood in the sense that, it maintains both growth and stability while reducing inequality is still far from being achieved. Which is all the more reason why one is at a loss how scholars like Song Byung-Nak rated Korea's economy as developed and even closing ranks with Japan, while both inequality and instability in the Korean economy were still deepening. The Korean second major economic crisis, the first having happened in the late 1970s which the state could not resolve, was yet another main proof of the fragility of the Korean economy.

If for anything, the recent crisis in the Korean economy is a very clear demonstration that the Korean economy was and still far from being rooted in the right economic fundamentals contrary to what scholars like Song Byung-Nak and multilateral financial institutions such as the World Bank and the IMF would want us to believe. The conditionalities for the IMF's bail-out have even shown clearly that, contrary to the propagation that the Korean economy was governed by the tenets of liberal economic theories, it is the state and not the market that has been propelling it. Part of this problem arises from the monogrowth strategy which the Korean state had adopted as a blueprint for economic development. This largely explains the difficulty the state has encountered in its atttempt to transcend the limits of mono-growth which the Korean economy has been trapped since the turn of the 1980s before the outbreak of the recent economic crisis in December 1997.

All this has far-reaching theoretical implications for the industrial and steel policies of the goverment of South Korean, which portends great lessons for Nigeria As noted earlier, in the late 1970s when the Korean economy was first faced with hyper-inflation, rather than revisit its dvelopment strategy with the hope of revising its industrial policy, the Korean government and some scholars of the Korean economy saw it as one of the problems expected to accompany a fast-growing economy. The majority of the Korean policy makers were, like then General Park, of the view that 'the cake should be baked first before sharing it'. What it all means, is that, the Korean government was not really concerned with a development strategy interested in promoting stability and equality, while the economy experienced growth. Suffice it to note, that the contemporary economic crisis in Korea signals that much was and is still very wrong with its economic development strategy.

The implications of all this for Nigeria, where the government is yet to formulate both industrial and steel policies is to avoid the Korean mistake by having an all-embracing development strategy that simultaneously promotes growth while ensuring stability of the economy and guaranteeing equity among the people. As part of the proposed development strategy, the Nigerian government should, rather than importing, use its huge steel raw materilas to develop the country's steel industry. That way, the state would have a large measure of control of the path and structure of the nation's industrialization.

Political System: Nothing has ever cast serious doubts on the efficacy of the political systems of South Korea to really sustain its economic growth much more bringing about economic development than the recent economic crisis. In fact, at a time when the debate is raging on among African scholars as to what type of political system should guide the continent into the next millennium, the economic crisis facing South Korea signals the limitations of political authoritarianism, serving as warning to most African heads of governments and policy makers who have endorsed it on the ground that it has worked well for the Asia NICs. Looking at the impressive growth of the 'Asian Tigers' economies whose political system was rooted in authoritarianism, the majority of African political leaders like the late head of the junta regime in Nigeria, Sani Abacha, not only began to fratenize with his counterparts in Asia, but also, started perfecting plans to adopt the same political system in Nigeria. Even before then, the impressive economic growth of the Asian NICs had, in the past, also triggered off heavy theoretical debate on the relationship between dictatorship and economic development in Africa, which was reactivated by the Uppsala-based Scandinnavian Institute for African Studies seminar held at Sweden in 1995. The main contention of a group of contributors to the seminar was that, on average, dictatorial civilian regimes as opposed to democratic governments in Africa, were able to promote policies that brought some reasonable level of economic growth. For instance, inspite of the authoritarian regime of Daniel Arap Moi, Kenya was seen by this group of scholars and multilateral financial institutions, as having sustained a measure of political stability that promoted growth in the Kenyan economy. Space will not permit us to revisit the whole debate here. Notwithstanding the kind of economic development that dictatorial regimes can bring about in the economy of a country, the contradictions it engenders, far out-weigh its benefits. In fact, the social and political upheavals that attend economic growth founded on autocratic regimes are so enormous that they cause a system break-down. Which is why Nigeria, or any other African country, should be cautious in adopting the industrial policies of South Korea and other Asian NICs.

As for the Korean government, it calls for a re-visit to the political system particularly along democratic lines since all previous regimes in Korea since 1961 have only succeeded in mimicing democracy. Democracy means more than conducting elections and having rule of law whereas the political system remains autocratic as has the case in the Republic. Until this fundamental political problem is resolved, the IMF bail-out measures can only be a stop-do-stop measure. For instance, President Kim Dae-Jung has, since assumption of office in January 1998, recognized the urgency of democratising the Korean state and its economy, having realised perhaps, that that is the only viable way which might bring the country back to its right footing. Kim's regime recent policy stance on Korea's political system represents a paradigm shift from the earlier political thought of the majority of Korea's political class and some scholars. Perhaps, this is due to the fact Kim is the first opposition candidate to be elected president of Korea since 1948, and wanted to inject a new thinking into Korea's polity. May be, it is all in the attempt to have Korea really put on the track of democracy. But the fear being raised by scholars is whether his regime has the political will to register the essence of real democratisation of Korea across all the institutional groups that make the Korean state a reality, and the Korean public. For, the majority of the Korean political class are opposed to a change of the current political system since it benefits them.

One major lessons from all the above analysis, is that, it would be misleading to recommend that Nigeria should adopt Korean industrial policy knowing fully well its origin and strength were rooted in an authoritarian political system. For, what Nigeria actually needs at the moment, after the military has been in government for more than 31 years since its political independence in 1960, is not an authoritarian regime, civilian or junta, as past experiences of the junta regimes in the country have shown that, it only helped the military and its cronies among the civilian political class, to prosecute vindictive politics, and degenerate the economy rather help bring about development. Nor would it be wise to endorse for Nigeria, a Korean type of state-led industrial growth that only enabled the state managers and their cronies to have firm grip of the commanding height of the Korean economy while deepening inequality among the mass of the Korean people. Rather, what one believes should be of main concern to policy makers and the majority of Nigerians who are interested in reversing Nigeria's political and economic backwardness, is how to dismantle all facets of authoritarian regime, military or civilian, and start afresh, the whole process of democratisation and the development of the country's economy.

Development Model: Since the World Bank's survey of the Asian NICs in 1993, some of the scholars writing on the emergent Asian economies as well as the majority of scholars of the African economy, have always mistakenly refered to them, the 'Asian Tigers', as a model. This is all the more so particularly when the impressive growth of the economies of the Asian NICs contrasted sharply with the poverty and debt-ridden economies of African countries. These were all mistaken positions as the basic unsettled theoretical issues that would facilitate a comparison of industrialiastion in Asia and Africa were not clarified in the first place. Instead, what was done in most cases, was to use indices like *per capita* income and gross national product to bring out the differences in the level of growth in the economies of Asian NICs and Africa. It is all wrong doing so because a development model presupposes that the Korean industrial policy, among others, is an ideal, whereas it is nothing of the sort.

Not only that, as it is pretty difficult to really talk of an ideal development strategy since even the approaches which some of the advanced industrial economies like Britain, the USA and Germany had adopted to industrialisation in the past, relatively successful they were, could hardly be regarded as a model given the differences in the historical experiences of these countries . It is perhaps for this and other related reasons that Korea's experience with industrial and steel development can hardly be replicated in Nigeria contrary to what the Bank and the Fund had suggested in the past. All this compels one to ask: What does

Song Byung-Nak mean by a 'Korean development model'? Why did he recommended it for other Third World countries including Nigeria to emulate?

Song had confused Korea's outstanding economic growth with a model, which it was and still is not. The fact that the Korean economy was and still obsessed with growth, has a fragile foundation, faced with deepening inequality, and operated an authoritarian political system and all of which made it to be far from being a model. In order to avoid the limits of the Korean development strategy, Nigeria should formulate industrial and steel policies that will take cognizance of its perculiar situation especially in terms of exploiting its raw materials for local use instead of importing them.

The fact that the Korean economy was far from being a model should not be contrued that the government did not make some level of progress in terms of industrial and steel development. Korea's economic growth has, no doubt, shown evidences that it was not really grounded in sound macro and micro-economic policies, making it all the more difficult for other developing economies like Nigeria to emulate. However, Korea's experience with industrial and steel development still presents some useful lessons both for the Korean policy makers and government officials, and for the government of other developing steel-producing country like Nigeria. For instance, inspite of the problems that were inherent in Korea's strategy of industrialisation, its rapid economic growth was partly indicative of the ability of an emergent state in the world system since the end of World War 11 when the North has continued to oppose the economic development of the South to use specific economic/industrial policy and actions to reverse the trend of underdevelopment which it had inherited from Japanese colonialism.

As noted earlier, using some major economic indicators like GNP and *per capita* income, South Korea was relatively not more developed than Nigeria by 1960. Yet, the Korean government was able to surmount some of the country's socio-economic problems such as a purely underdeveloped agrarian economy, industrial underdevelopment, illiteracy, poverty, lack of social amenities, and brought significant economic growth within two decades. Not only that, without iron ore, coal and alloying minerals, Korea rose to become the sixth world largest producer of crude steel in 1995. And much more importantly, Korea used its steel sector to launch its industrialisation irrespective of the limitations of the mono-growth strategy of the Korean government development plan. Viewed against the woeful attempt at steel development by the Nigerian government, our concern at this stage of the work, is to pinpoint the nature of the overall lessons that ought to have been learnt by Nigeria from the Korean experience and, in turn, how they would influence recommending policy for Nigeria's steel industry.

7.4 WHAT SHOULD BE DONE TO DEVELOP NIGERIA'S STEEL INDUSTRY: SOME POLICY RECOMMENDATIONS

7.4.1 What The Goals of Industrial And Steel Policies Should Be In Nigeria:

As noted earlier, the path that the Nigerian state took to industrialisation gave useful insights into why the steel sector has failed to play the role of a catalyst for Nigeria's industrialization. The absence of an industrial policy to guide steel development made its failure inevitable. In other words, the policy statements that have been made by successive governments in Nigeria on the urgent need for the industrial transformation of the country, were merely defensive radical statements pretending as if industrialisation was on the agenda of those in charge of the state, whereas it is not.

Yet, Nigeria has got virtually all that it would take to get the country industrialised. For instance, Nigeria is endowed with almost all the known industrial minerals including oil and gas. Also, Nigeria has large acres of arable land for the agri-business industry to flourish. Korea's experience in this context, especially as a country lacking basic minerals for industrialization, is quite instructive. For instance, Korea was faced with the reality that, unless it industrialises itself, no other country will do it. Not even the USA which helped restore some measure of political stability in Korea after the 1950-1953 Korean war was really interested in South Korea's industrial development. Suffice it to say that, the task of getting Nigeria's industrialization really underway, is basically that of the government, people, and the private investors. The goal of Nigeria's quest for industrialisation should be the creation of a self-reliant, equitable, and balanced economy that is anchored on a sound foundation of light, intermediate and capital goods production, which, in turn, are organically interlinked with other sectors of the economy, particularly mining and agriculture. This was quite a noble goal, but the pertinent question is whether it be can achieved.

Furthermore, the Nigerian state had embarked on steel development without any policy as explained earlier. This was not surprising since steel development in the first place was not really on the agenda of those in charge of the state. If for anything, government's intervention into the steel sector not only came at a time when the country was awashed with wealth from oil production and export, but also, when the state managers wanted yet another avenue for sustaining primitive accumulation. Not unsurprisingly, there was still no steel policy formulated by the government as of 1998. Rather, what has been done by the Nigerian government was to ask the Ministry of Power and Steel to appraise the ailing public steel companies and then recommend ways out of the crisis. Perhaps, it is from the recommendations of the Ministry and the special committees such as the one recently set up by the head of the junta, General Abubakar that a policy on steel might emerge. Granted that it was effort in the right direction, it was still focused on the public steel sector, to the detriment of the equally crisis-ridden private steel scetor. And that is itself, problematic. In essence, the Nigerian government was yet to be well focused on its steel development project at the time this study was conducted.

In order to reverse Nigeria's industrial underdevelopment, the goal of the steel policy within the overall industrial policy of the Nigerian state, should be the promotion of self-sufficiency in steel raw materials, and for the steel companies to act as a catalyst for the country's industrialisation. As part of the broad steel policy, the huge local deposits of steel raw materials in the country, should be exploited for use by the steel companies. It should be made illegal to import iron ore, coal and the alloying minerals which are found in mineable large quantities and of good quality in the country. The integrated steel companies should have production lines where flat steel products and other capital goods, are rolled out. In that way, the product mix of the integrated steel companies would serve as inputs for other intermediate and capital goods producers while helping to reshape the structure of manufacturing in Nigeria's economy which has been no less than mere assembly of capital goods, and bedevilled by high import dependency ratio. Korea's experience in this regard is quite instructive for Nigeria. For, the relative success of Korea's steel sector in comparison with Nigeria, lies largely in the wide product range of POSCO which enabled it to quicken the country's industrialisation. Now let us return to what should be done to achieve these goals.

7.4.2 Suggested Alternative Plans of Action:

Domestic Environment: Within the domestic terrain, the Nigerian state and its component groups have to be restructured. This is because the state is not only central to the process of industrial accumulation in Nigeria, but also, those in charge of the state would, in an attempt to consolidate their means of survival, oppose any change of the existing political power configuration and the mode of surplus extraction in the country. With substantive democratic means and not just following procedure, the Nigerian state can be restructured in such a way that its proposed industrial policies and projects are concerned with the improvement of the material well-being of the people and the economy instead of serving parochial interests. The process of restructuring the state should begin with the replacement by democratic means, of all those who are currently controlling it and across its institutional groups by elected people with proven intergrity, who, in turn, should be made to mandatorily account to the people. This is not to say that, with a mere return to civilian rule in Nigeria, all will be well with the Nigerian state. Far from it. For, that, indeed, will mark the beginning of an arduous task of restucturing the state as those who benefit from the existing structure will still be opposed to such a change. As part of the process of restructuring the state, too, those expected to manage it and its apparatuses, should all undergo a re-orientation as to the new ideology of what politics should be. That is, as a game, but not as a means to the accumulation of wealth. They should also be made to understand that, public policy including the state's policy on industrialisation should be people-oriented, but not to serve the parochial interests of those in charge of the state. This is particularly so since man and his material well-being should constitute the essence of economic and political development, and any other development for that matter.

Perhaps, with the restructuring of the Nigerian state in line with the suggestions made above, it would become a lot easier for its managers to formulate an industrial policy which they should have the political commitment to implement. Rather than continue with the local reproduction of hitherto imported intermediate and capital goods, the industrial policy being proposed should enable the state to surmount the obstacles placed on its path by foreign capital to the local substitution of these goods. One practical step towards achieving this goal, is for the integrated steel companies to be fitted with production facilities that would roll out intermediate and capital goods either as inputs for other capital goods producers, or as finished goods. That way, technology acquisition and accumulation would be developed locally in the country.

The industrial policy should not make the country's upstream steel sector the exclusive preserve of the Nigerian state, but allow the private investors entry into it. As noted, South Korea's poverty in iron ore and coal, and the resultant high import dependency profile, compelled the government to discourage private local capital from intervening into the upstream sector of its steel industry. However, that is not the situation in Nigeria, where virtually all the steel raw materials abound. In essence, the industrial policy of the Nigerian state should be able to coordinate and encourage both public and private steel companies to engage in the upstream sector of the country's steel industry with strong emphasis on supplementing each other's effort. One of the advantages of such approach is that, it would permit a comprehensive economic development in which there will be complimentarity between and among the light, intermediate and capital goods producing companies, and between them and the rest of the economy. Also, the situation in the past when both public and private manufacturing companies had to import most of their basic production inputs would be avoided in the future.

Conclusion

On Nigeria's specific steel policy, the country's steel sector should be made to really launch its industrialisation and technological breakthrough. Achieving this goal would require that, the steel companies should have a wider range of products from which other light, intermediate and capital goods producers can source their production inputs. Once this is done, it will go a long way to correct the mistakes arising from Nigeria's previous quest for industrialisation. Within the broad framework of the steel policy also, there should be an incremental reduction of imported steel products over an agreed period of time by having them produced locally until such a time that, full self-sufficiency is achived. This would help check the dumping of sub-standard steel products into the country thereby avoiding the frequent cases of buildings collapsing either during construction, or shortly after completion.

Within this broad context of steel policy, too, the Nigerian state should not go beyond the stage of late growth in the country's steel industry while helping to develop the private steel sector. This is to avoid the crisis in which the Korean steel industry is currently faced with: a crisis of the inability of the Korean state to move the country's steel industry from late growth into the stage of market maturity. For, the politics of the state and its mode of capitalism are incompatible with the principles of the market. So, state's dominance of the Korean steel industry constrained both the activities of local private capital and market forces from flourishing in the industry.

What should the role of the Nigerian state be in the steel sector? Given the past experiences of the major steel producing countries of the world like Germany, France and Japan, the state had played an active role in the steel sector right from the stage of take-off through early growth to late growth. This was because of the political and social implications of steel development at those levels. Thereafter, the state did not only divest from, but played a regulatory role in the steel sector, but left the future operations of the industry which should be based on market mechanisms, to the private steel companies. With a restructured Nigerian state as proposed above, it should, after taking the country's steel sector right from the take-off stage to late growth, delink significantly from the steel industry and get itself concerned with playing largely a regulatory role in the industry. As explained earlier, at late growth, the state should have accomplished the social roles that prompted its intervention in steel development. As part of its regulatory role, therefore, the state should create an enabling political and economic environment that would permit the major private steel companies to continue with the efficient operations of the steel industry. It is the private steel companies with the state playing a supervisory role that will then move the country's steel industry into market maturity.

With the right managers in place, the Nigerian state should still remain largely as an active investor in the country's steel industry particularly as it is still at the take-off stage. This is contrary to the pressure mounted by the World Bank and the IMF on the Nigerian government to commercialise the public steel companies. The argument of the Bank and Fund as stated earlier, has always been that, the state is a 'bad manager' of investment including public companies and therefore, should have no business in the steel sector beyond playing a regulatory role. But one of the major reasons why Nigeria's steel industry is beset with crisis right from its inception, is not so much with involvement of the state as its custodians' own conception of industrialisation and steel development. So, the real problem is not that the state intervened in steel development as that has been the first step even by the major steel producers of today; nor was it really a bad manager. Rather, it is the state managers' own concept of steel development which is an off-shoot of its thinking of industrialisation. For, if the state was able to get underway, the steel industries of France, Germany and Japan among other major world steel producing countries, then, the way out of the crisis for Nigeria is to have the state managers re-oriented towards ensuring that the steel industry quickens the country's industrialisation. Once this is done, steel development should be placed on the agenda and cease to be a political project. This would mean that, the Nigerian government and people should be well focused and committed to steel development.

External Environment:

At the external level, the major policy areas that one considers necessary for Nigeria's industrialisation generally and steel development in particular, are the need for foreign investment code, the role Nigeria should play in the steel industry of West Africa, and bilateral relationship with the industrial economies.

Conclusion

The Nigerian state should formulate an investment code to regulate the inflow of foreign capital into Nigeria's economy. This is because a lot of what has been taken as an investment code were no less than incentives to attract foreign investors into the country. Far from being an error, it was all part of the conscious plan of the managers of the Nigerian state to inherit, but not to change the exploitative industrial and economic policies of the colonial state. This is all the reason why the absence of a well articulated foreign investment code had largely accounted for the failure of Nigeria's efforts at industrialisation generally and steel development in particular. Due in part to the lack of an investment code in Nigeria, the state has been unable to really direct foreign capital into the sector of the economy it wanted. It is for the same reason that the state could not formulate policies that should protect the infant industries from the stiff opposition posed by foreign capital over the control of the post-colonial economy of Nigeria. In the light of all this, the policy content of the investment code being recommended for Nigeria should first re-prioritize the type of industries to be established in the country, taking cognizance of the choice of technology to be used. All cottage industries should be made the exclusive preserve of local private capital since they serve as early foundations for local acquisition and acummulation of technology in the country. Other categories of industries which should be made the exclusive preserve of local private capital were already contained in the Indigenization Decree of 1977 before then Abacha regime later repealed it. As part of the investment code, these categories of industries should be revisited and be made exclusive for local private capital. Foreign capital was able to cripple the indigenization decree partly because it still controlled the importation of CKDs thereby reducing the local content of manufacturing. The indifference of the state managers to the decree made it all the more so. The Korean experience in this context is quite instructive. For, one of the reasons for the relative success of Korea's policy on foreign investment especially in protecting the infant industries, was because the state was committed to it. Foreign investment policy should be reviewed over a specific period in order to avoid the Korean mistake of prolonged closure of its economy to foreign capital, which prompted the OECD's pressure on the government of the

Republic of Korea to open up its economy for exports and investors from the member-countries of the Organisation and the West. The pressure is being reinforced by the IMF that has asked the Kim Dae-Jung regime to open the relatively closed Korean economy to the West.

As part of the policy content and direction of the investment code, the Nigerian government should regulate the pattern and sectoral flow of all foreign investment particularly into the industrial sector of the country. The code should spell out the details of the technological content of any assembly or manufacturing of intermediate and capital goods that foreign capital is involved either with the state, or in partnership with local private capital. This is aimed at avoiding the situation in the past which still persist at the time this study was conducted, whereby the Nigerian government in particular, and the economy at large were nothing but mere consumers of imported intermediated and capital goods, and the foreign technology. In other words, the investment code being proposed here should emphasize the development of local technical capacity to replicate technology by protecting the infant industries. The Andean Pact's Decision 24, too, otherwise known as the Investment Code for its member-nations, is quite instructive for Nigeria. For, the Pact's investment code had far-reaching implications in the sense that it was aimed at controlling the inflow of foreign capital in the developing economies of the Andean sub-region, especially as it affects employment, duplication of the activities of existng industries, technology, and dependence. What it all portends for Nigeria in terms of the mix of its foreign investment code, is that, rather than allowing foreign capital an unfettered entry into and exit from, Nigeria's economy generally, the investment code should effectively tackle the fundamental issues like dependence, job creation, and technology acquisition and accumulation locally.

Nigeria should undertake an export-oriented industrialisation in West Africa. This, it could achieve by taking the lead in the promotion of industrialisation and steel development within the sub-region. For, out of the 10 iron and steel producing countries in Sub-Saharan Africa, only Nigeria and Zimbabwe have integrated steel companies inaddition to many steel mini-mills. The remaining eight countries notably Kenya, Ghana, Tanzania and Cote d'Ivoire, their steel sectors were all dominated by mini and rolling mills of installed annual capacities not exceeding 50,000 metric tonnes of crude steel. Other countries like Liberia and Mauritania produce iron ore almost wholly for export. In other words, Sub-Saharan Africa's steel industry is, on average, still largely youthful with Nigeria accounting for most of the activities in the industry in the continent, inspite of the problem its industry is faced with. With its vast untapped steel raw materials, size and capacity of its steel companies, Nigeria remains the most viable steel producer both in the West Africa sub-region and in Sub-Saharan Africa as a whole, if all had gone well with steel development in the country.

One of the major reasons why the economies of the member-countries of the ECOWAS have still remained underdeveloped is because of the inability of these countries, to harmonize their micro and macro-economic policies, particularly in their industrial sector. In light of Nigeria's vast deposit of steel raw materials and other industrial minerals, it has the potential of becoming a major steel producer in Africa if all goes well with steel development in the country. Not only that. Nigeria could use its steel sector to push an export-oriented industrial policy within the sub-region in a way similar to what South Korea did with its steel sector in the East Asian countries. As part of this policy, Nigeria should use the framework of ECOWAS to restructure the steel sectors of its member-countries. The restructuring of the sub-region's steel industry would require that, not all countries in the region should really have an integrated steel company. Rather, with the proper co-ordination of the industrialisation projects under the aegis of ECOWAS, Nigeria's steel industry should be made to service the steel sectors of smaller steel-producing countries like Ghana and Cote d'Ivoire whose major steel companies were merely mini and rolling mills. For countries whose economies have very low absorptive capacity for steel products such as Togo and Niger, they should be made to have only rolling mills. Again, with the harmonisation of its fiscal and economic policies, Nigeria could also use the framework of ECOWAS to re-direct the use of a larger chunk of the iron ore produced in Liberia and Mauritania which was all exported to Europe, back to its integrated steel plants. The initial step which the Nigeria government took by investing in the Mifergui/Nimba iron ore deposit in Guinea and later pulled out, should be revisited with the hope of getting it underway. For, it would further beef up the country's steel raw material base and see it into becoming an industrial giant not only within the sub-region, but in the developing economies.

Third and final policy area is that, Nigeria's bilateral relationship with the donor countries should shift from financial aids and loans to capacity building in manpower and technology. For too long, the nature of Nigeria's relationship with bilateral donor-countries has been largely in the form of aids either in cash or in kind. However, the politics of these aids has shown that they represent no less than an attempt to deal with the major market problems which the donor-countries have been trying to resolve, but with limitd success. The politics of aids makes it hardly an effective weapon for dealing decisively with the long-term development goals of the recipient country. The same is also true of bilateral loans. In most cases, the lending country converts loans into goods, which it also supplies to the debtor country thereby leaving it worse-off. For instance, Kobe Steel of Japan had, in the past, advanced a US$1.6million loan to the Katsina Steel Rolling Company; but the loan was later converted into billets, which it [Kobe Steel] also supplied. In essence, it was no less than using aid as another marketing strategy, a practice that has dominated the flow of aids from the North to the South since the turn of the 1960s when foreign capital began to face problems of overconcentration and declining returns on investment among others. For Nigeria to avoid such situation in the future and derive the greatest possible gains from its bilateral aid relations or financial loan from other countries, the priorities should be such that would promote the long-term goals of Nigeria's industrialisation policy and projects.

7.5 CONCLUSION

Nigeria has huge untapped industrial and steel raw materials which if exploited and effectively used, can transform the country into an industrial giant both in Africa and the developing economies at large. The Korean recent economic crisis is a pointer to the danger of the over-presence of the state in the real sector of the economy of developing countries. But it also cautions against abandoning the economy wholly to the forces of the market. In order to realise the goal of Nigeria becoming a

Conclusion

major steel producer and an industrial power in the developing world, the country should have a stable domestic political enviroment and the right economic policies and mixes. This should be part of Nigeria's challenges in the 21st century, particularly as it stirves to reposition itself in the global economy. This is no mean task ahead. And for Nigeria to succeed, it requires the full commitment of all: the state, local private capital and the people.

NOTES AND REFERENCES

1. See F. Deyo, *The Political Economy of the New Asian Industrialism*, *op. cit.*
2. R. Kiely, 'Development Theory and Industrialisation: Beyond the Impasse', *op. cit.*
3. M. Feldstein, 'Refocusing the IMF', *Foreign Affairs*, March/April 1998.
4. D.R. Henderson, 'Secrets of East Asia's Tiger Economies', *Orbis*, Summer, 1997.
5. D. Brautigam, 'What can Africa Learn from Taiwan?: Political Economy, Industrial Policy, and Adjustment', *Journal of Modern African Studies*, 32, 1, 1994.
6. S. Radelet et al, 'Asia's Bright Future', *Foreign Affairs*, November/December, 1997.
7. *ibid.*

Bibliography

Adu, A. Boahan, 1985: *Africa Under Colonial Domination: General History of Africa*, vol.ii, Heinemann and UNESCO.
Ahn, Byung-joon, 1998: 'Prospects for Korea Under the IMF and Kim Daejung', *Korea Focus*, May-June, 6.3.
Ake, C., 1978: *Revolutionary Pressures in Africa*, London, Zed Press.
————, ed., 1985: *Political Economy of Nigeria*, London, Longman.
————, 1996: *Democracy and Development in Africa*, Washington D.C., The Brooking Institute.
Akerele, A., 1990: 'Nigeria's Patent System and Technology Development Strategy', in Ige, C. ed. *Capital Goods and Technological Development in Nigeria*, Minna, Economic Society Conference Proceeding.
Alam, S.M., 1989: *Government and Markets in Economic Development Strategies: Lesson from Korea, Taiwan and Japan*, New York, Praeger Publishers.
Alt, James E. and Kenneth, A. Shepsle, eds., 1990: *Perspectives on Positive Political Economy*, New York, Cambridge University Press.
Amsden, H. Alice, 1989: *Asia's Next Giant: South Korea and Late Industrialization*, New York, Oxford University Press.
Amsden, Alice and Linsu Kim, 1986: 'A Technological Perspective on the General Machinery Industry in the Republic of Korea', in *Machinery and Economic Development*, ed. Martin Fransman, London, Macmillan Press.
Andrae, G. and Beckman, B., 1987: *Industry Goes Farming: The Nigerian Raw Material Crisis and The Case of Textile and Cotton*, Uppsala.
Annual Abstract of Statistics 1989-1997: Lagos, FOS.
Atlas of African Industry: Iron and Steel, 1989: Vienna, UNIDO.
Ayodele, A., 1990: 'Nigeria's Technology Issues - A Revisitation', in Ige, C. ed. *Capital Goods ...*
Ayoola, G., 1990: 'Technological Progress in Agriculture: Some Issues', in Ige, C. ed. *Capital Goods ...*
Balassa, Bela, 1981: *The Newly Industrialized Countries in The World Economy*, New York: Pergamon Press.
————, ed. 1982: *Development Strategies in Semi-Industrial Economies*, Baltimore, John Hopkins University Press.
Bangura, Y., Mustapha, R. and Adamu, S., 1983: 'The Deepening Economic Crisis and Its Political Implication', Paper submitted to the Zaria National Conference on The State of the Nation.

Bangura, Y., 1987: 'Adjustment and De-Industrialization in Nigeria', mimeo.
———, 1989: 'Crisis Adjustment and Politics in Nigeria', *Akut*, 38 Uppsala.
Baran, Paul, 1957: *The Political Economy of Growth*, New York, MRP.
Blomstrom, M. and Hettne, B., 1984: *Development Theory in Transition: The Dependency Debate and Beyond: Third World Response*, London, Zed.
Bonazza, Patrick, 1978: *Iron Ore and Europe*, Brussels, European News Agency.
Brautigam, A. Deborah, 1994: 'What Can Africa Learn From Taiwan?: Political Economy, Industrial Policy and Adjustment', *Journal of Modern African Studies*, 32.1.
Brown, M., Damment, A., Meeraus, A. and Stoutjesdijic, A., *Worldwide Investment Analysis: The Case of Aluminium*, World Bank Working Papers, 603.
Bruneau, Thomas, C., and Philippe Faucher, eds. 1981: *Authoritarian Capitalism: Brazil's Contemporary Economic and Political Development*, Boulder, Westview Press.
Cardoso, Fernando, H. and Enzo, Faliho, 1979: *Dependency and Development in Latin America*, Berkeley, University of California Press.
Cho, Soon, 1994: The *Dynamics of Korean Economic Development*, Washington D.C., Institute for International Economics.
Cummings, Bruce, 1984: 'The Origins and Development of The Northeast Asian Political Economy: Industrial Sectors, Product Cycles and Political Consequences', *International Organisation*, vol.38, no.1, Winter.
Deyo, C. Frederic, ed. 1987: *The Political Economy of The New Asian Industrialism*, Ithaca, Cornell University Press.
Eboji, P., 1990: 'The Military and Technology Development in Nigeria', in Ige, C. ed. *Capital Goods* ...
Ejembi, H., 1982: 'The Current Austerity Measures and Their Implication for the The Working People of Nigeria', (Mimeo).
Ekuerhare, B., 1985: 'On Industrial Underdevelopment in Nigeria: A Theoretical Celebration of Walter Rodney', *African Development*, x, iv.
———, 1986: Pattern of Manufacturing, Industrial Growth and Import Constraints in Nigeria', *West African Economic Journal*, vol.6.
———, 1990: Second Tier Foreign Exchange Market and Structure of Industrial Growth in Nigeria', in Olaniyan, R. and Nwoke, C. eds. *Structural Adjustment in Nigeria: The Impact of SFEM on the Economy*, Lagos, NIIA.
———, 1990: The Theory and Concepts of Capital Goods in Nigeria', in Ige, C. ed. *Capital Goods* ...
———, 1996: *Pattern and Problems of Industrial Accumulation in Nigeria*, Lagos, Vanhurst
Ekundare, O., 1973: *An Economic History of Nigeria: 1860-1960*, London,

Methuen & Co
Engels, Fredrick, 'The Origin of the Family, Private Property, and the State', in Selected Works, ed. *Marx and Engels*, vol. ii.
Erlich, A., 1978: 'Dobb Marx Feldman Model: A Problem in Soviet Economic Strategy', *Cambridge Journal of Economics*, 2.2.
Evans, Peters, B., 1979: *Dependent Development: The Alliance of Multi-National, State and Local Capital in Brazil*, Princeton, Princeton University Press.
——————, 1987: 'Class, State and Dependence in East Asia: Lessons for Latin Americanists', in Frederic Deyo ed., *The Political Economy of The New Asian Industrialization*, Ithaca, Cornell University Press.
Fadahunsi, A. and Igwe, B.U.N. eds., 1989: *Capital Goods, Technological Change and Accumulation in Nigeria*, Dakar, CODESRIA.
Fadahunsi, A., 1989: 'Agricultural Machines, Equipment and Other Agro-Based Industries', in Fadahunsi and Igwe, eds., *Capital Goods, Technological Change ...*
Feasibility Report on The Proposed Ajaokuta Steel Company, 1978: Lagos, Ministry of Mines, Power and Steel.
Feldman, 1972: Feldman Model as presented in E. Domar: A Soviet Model of Growth', in Nove, A. and Nuti, D.M., *Socialist Economics*, Harmondsworth, Peguine.
Feldstein, Martin, 1998: 'Refocusing The IMF', *Foreign Affairs*, March/April, 77, 2.
Frank, G. Andre, 1967: *Capitalism and Underdevelopment in Latin America: Historical Studies of Chile and Brazil*, New York, Monthly Review Press.
——————, 1969: *Latin America: Underdevelopment or Revolution*, New York, MRP.
——————, 1977: *Imperialism and Unequal Development*, New York, MRP.
——————, 1979: *Dependent Accumulation and Underdevelopment*, New York, MRP.
Fransman, Martin and Kenneth, King, 1988: *Technological Capability in The Third World*, New York, St. Martins Press.
Friedman, David, 1988: *The Misunderstood Miracle: Industrial Development and Political Change in Japan*, Ithaca, Cornell University Press.
Gereffi, Gary and Donald, L. Wyman, eds., 1990: *Manufacturing Miracles: Paths of Industrialization in Latin America and East Asia*, Princeton, Princeton University Press.
Graham, R., 1982: *The Aluminium Industry and The Third World: Multinational Corporation and Underdevelopment*, London, Zed.
Green, David, 1967: *Steel and Economic Development: Capital-Output Ratio in Three Latin American Steel Plants*, Michigan, Michigan State University Press.
Han, Sung-joo, 1991: 'The Korean Experiment', *Journal of Democracy*, 2.2
Harold, Peter, Jayawickrama Malathi and Bhattasali Deepak, 1996: 'Practical Lessons for Africa from East Asia' in Industrial and

Trade Policies, World Bank, (Discussion Papers).

Hatch Report, 1988: On Steel Sub-Sector in Nigeria, Canada, Hatch Associates (World Bank Consultant).

Heavy and Chemical Industry Promotion Committee (HCIPC), 1973: The Reorganization of Industrial Structure Based on HCI Plan, Seoul, HCIPC.

——————, 1976: Heavy and Chemical Industry, Seoul, HCIPC.

Henderson, R. David, 1997: 'Secret of East Asian's Tiger Economies', *Orbis*: Summer.

Hinton, Harold, C., 1983: *Korea Under New Leadership: The Fifth Republic*, New York, Praeger.

Hirschman, Albert, O., 1968: 'The Political Economy of Import-Substituting Industrialization in Latin America' *The Quarterly Journal of Economics*, No.82 (February).

Hoffman, W.G., 1958: *The Growth of Industrial Economics*, Manchester, Manchester University Press.

Hogan, William T., 1983: *World Steel in The 1980s: A Case of Survival*, Lexington, Lexington Books'

——————, 1985: 'POSCO: Continues to Grow', *Iron and Steel Engineer*, 62, 4. April.

——————, 1991: *Global Steel in The 1990s: Growth or Decline?*, Washington D.C., Heath and Company.

Howell Thomas, R. ed., 1983: *Steel and The State*, Boulder, Westview Press.

Huntington, Samuel and Myron Weiner, 1987: *Understanding Political Development*, Boston, Little Brown and Company.

Huntington, Samuel, 1968: *Political Order in a Changing Society*, New Haven, Yale University Press.

Igwe, B.U.N. 1989: 'The Iron and Steel and Machine Tools', in Fadahunsi and Igwe eds., *Capital Goods and Technological Change* ...

Inuwa, I.K., 1990: 'The State of Technology Development in Nigeria', in Ige, C. ed., *Capital Goods* ....

Iniodu, R.U. and Ukpon, I.I., 1990: 'Capital Goods and Technological Development in Nigeria Agricultural', In Ige, C. ed., *Capital Goods* ...

Jacobs, Norman, 1985: *The Korean Road to Modernization and Development*, Chicago, Chicago University Press.

Jacquenim, Alexis, 1987: *The New Industrial Organization: Market Forces and Strategic Behaviour*, Cambridge, MIT Press.

Jei, Guk-jern, 1995: 'Exploring The Three Varieties of East Asia's State-Guided Development Model: Korea, Singapore and Taiwan', *Studies in Comparative International Development*, Fall, 30.3.

Johnson, W., 1966: *The Steel Industry of India*, Mass, Harvard University Press

Jones, Kent, 1986: *Politics Vs Economics in World Steel Trade*, Boston, Allen & Unwin.

Jones, Leroy, P., 1980: *Jae-Bul and The Concentration of Economic Power in Korean Development*, Seoul, KDI.

Jones, Leroy, P. and Il Sakong, 1980: *Government Business and*

Enterpreneurship in Economic Development:The Korean Case, Cambridge, Havard University Press.'

Kayode, M.O., 1987:'The Structural Adjustment Programme and The Industrial Sector', in Philips, A. and Ndekwu, E. eds.,*Structural Adjustment in a Developing Economy:The Case of Nigeria*, Ibadan, Ibadan University Press.

KDI, 1985: *High Technology Industry Structure and Reality*, Seoul, KDI Press.

KEB, 1980:'Adjustment of Korea's Heavy Chemical Industry',*Monthly Review*, December.

——————, 1985:'High Technology Industries in Korea',*Monthly Review*, May.

——————, 1987:'The Iron and Steel Industry in Korea',*Monthly Review*, February.

Keily, Ray, 1994:'Development Theory and Industrialization: BeyondThe Impasse',*Journal of Contemporary Asia*, 24,2.

Kendrick, D., Meeraus, A. and Alatome, J., 1984:The Planning of Investment Programme in The Steel Industry,Vol.3, World Bank, John Hopkins University Press.

Kilby, P., 1969: *Industrialization in an Open Economy: Nigeria 1945- 1966*, Cambridge, Cambridge University Press.

Kim, Dong-ki and Linsu Kim, 1989: *Management Behind Industrialization: Reading in Korean Business*, Seoul, Korea University Press.

Kim, Ilpyong, J. and Young Whan Kihl, eds., 1988: *Political Change in South Korea*, New York, Korean PWPA, Inc.

Kim, Kyung-Dong, ed., 1987: *Dependency Issues in Korean Development: Comparative Perspective*, Seoul, Seoul National University Press.

Kim Kyung-Dong, 1976:'Political Factors in The Formation of the Enterpreneurial Elites in South Korea',*Asian Survey,Vol.xiv, No.5*.

Kim, Linsu, 1989: Science and Technology Policies for Industrialization in Korea', in *Strategies for Industrial Development*, edited by Jangwn Suh, Kuala Lumpur, Asia and Pacific Development Council.

Kim, Suk-joon, 1989: *The State, Public Policy and NIC Development*, Seoul, Daeyoung Moonwhasa.

Kolade, B., 1990:'Capital Goods Development in Nigeria', in Ige,C. ed., Capital Goods

Korea Development Bank (KDB), 1986: *Industry in Korea*, Seoul, KDB.

Korea Exchange Bank, 1979:'The Iron and Steel Industry in Korea', *Monthly Review*, February.

Korea Iron and Steel Association (KISA), 1987: Chulgang Tongkye Nyunbo (Iron and Steel Statistical Yearbook), Seoul, KISA.

Krugman, Paul, 1994:"The Myth of Asia's Miracles",*ForeignAffairs*, November/December.

——————, 1997:'Is Capitalism Too Productive?',*Foreign Affairs*, 75.6.

Kuk, Minho, 1988:'The Governmental Role in The Making of Chaebol in The Industrial Development of South Korea',*Asian Perspective*, December.

Kuznet, Paul, W., 1977: *Economic Structure in the Republic of Korea*, New Haven, Yale University Press.

Kyang-Hwie, Mihn, 1988: *Industrial Policy for Industrialization of Korea*, Seoul, KIET.

Lau, Lawrence, J. ed., 1990: *Models of Development: A Comparative Study of Economic Growth in South Korea and Taiwan*, San Fransisco, ICS Press.

Lee, Chung-Hu, 1991: 'The Government and Financial System in The Economic Development of Korea', *World Development*.

Lee, Kyu-Uck, 1986: *Industrial Development: Policies and Issues*, Seoul, KDI Press.

Lewis, A., 1955: *The Theory of Economic Growth*, London, Allen & Unwin.

Lim, Hyun-Chin, 1985: *Dependent Development in Korea*, Seoul, Seoul National University Press.

Lister, Louis, 1960: *Europe's Coal and Steel Community: An Experiment in Economic Union*, New York, Twentieth Century Fund.

Mamdani, M., 1985: 'Disaster Prevention: Defining The Problems', (Mimeo).

Manufacturers Assocition of Nigerian (MAN) Half-Yearly Review, 1981-1996, Lagos.

———, 1990: *Nigeria Industrial Directory*, Lagos.

Marcus, Peter, F. and Karlis, M. Kirsis, 1985: *POSCO: Korea's Emerging Steel Giant*, New York, Paine Webber.

Marx, K., 1978: *Capital: A Critique of Political Economy*, Vol. II.

McGannon, Harold, F. ed., 1971: *The Making, Shaping and Treating of Steel*, Pittsburgh: Herbick & Held.

Metal Bulletin, 1975-1995 Issues.

Miliband, Ralph, 1969: *The State in Capitalist Society*, New York, Basic Books.

MITI, Japan 1988: 'The Role of Industrial Policy in The Post WorldWar II Economic Development in Japan', (Mimeo).

Mustapha, R., 1983: 'Repression and The Nigerian Political Economy' (Mimeo).

Myrdal, Gunnar, 1968: *Asian Drama: An Inquiry into the Poverty of Nations*, New York, Pantheon Press.

Nigerian Economic Society, 1988: Structural Adjustment Programme and The Nigerian Economy Conference Proceedings

———, 1990: Capital Goods and Technological Development in Nigeria, Minna.

Nigerian Metallurgical Society, 1994: The Strategic Potentials of the Aluminium Smelting Industry to the Nigerian Economy, Lagos (June), Seminar Proceedings.

———, 1994: The Nigerian Steel Industry: Techno-Economic Appraisal, Lagos (November), Seminar Proceedings.

Nnoli, Okwudiba, ed., 1981: *Path to Nigeria's Development*, Dakar, CODESRIA.

Odama Report, 1983: Report of the National Economic Council Expert

Bibliography

Committee on the State of the Nation, Lagos, Government Printers.

Ogwumike, F.O. and Emenuga, C.E., 1987: *Review and Appraisal of SAP,* Lagos, Government Printers.

Ohiorhehenuan, J., 1987: 'Re- Colonizing Nigerian Industry: The First Year of the Structural Adjustment', in Philip and Ndekwu eds. *Structural Adjustment ...*

Okimoto Daniel, I, 1989: *Between MITI and The Market: Japanese Industrial Policy for High Technology,* Stanford, Stanford University Press.

Olsen Edward, A., 1980: 'Korea Inc: The Political Impact of Park Chung-Hee's Economic Miracles', *Orbis,* Vol.20, No.1 (Spring).

Olukoshi, A., 1986: 'The Multinational Corporation and Industrialization in Nigeria: A Case Study of Kano C. 1903-1985', Unpublished Ph.D Thesis, Leeds University.

Olukoshi, A., ed. 1991: *Crisis and Adjustment in the Nigerian Economy,* Lagos, Jad.

Olukoshi, A., Adewusi, A., Obi, C., Omoweh, D., and Bala, J. 1995: *Industrialization in Nigeria: Problems and Prospects,* Tokyo, IDE.

Olukoshi, A., 1991: 'The Performance of Nigerian Industry Under the Structural Adjustment Programme', in Olukoshi, A. ed., *Crisis and Adjustment...*

—————, 1991: 'An Assessment of The Impact of The Economic Recovery Programme of the Nigerian State on The Manufacturing Sector: A Kano Case-Study', (Monograph), Social Science Council of Nigeria.

————— 1993: *The Politics of Structural Adjustment in Nigeria,* London: James Currey.

Omoweh, D, 1996, 'The Nigerian Steel Sector in the Global Steel Industry', *Annals,* Social Science Council of Nigeria.

—————, 1991: 'Structural Adjustment and The Nigerian Iron and Steel Industries', in Olukoshi, A. ed, *Crisis and Adjustment ....*

Omoweh Independent Committee Report on the Commercialisation of Katsina Steel Rolling Company, 1992: Commissioned Study by The Katsina Steel Rolling Company

Onimode, B., 1985: 'The Political Economy of Capital Goods Manufacturing in Nigeria', in Fadahunsi and Igwe eds., *Capital Goods ...*

Palma, Gabriel, 1978: 'Dependency: A Formal Theory of Underdevelopment or Methodology for The Analysis of Concrete Situations of Underdevelopment', *World Development, No.6.*

Park, Chung-hee, 1971: *To Build a Nation,* Washington D.C., Acropolis Books Limited.

—————, 1971: *Our Nation's Path,* Seoul, Hollym Publications.

—————, 1970: *The Country: The Revolution and I,* Seoul, Hollym Publications.

Phillips, A. and Ndekwu, E. eds., 1989: *Structural Adjust Programme in a Developing Economy: The Case-Study of Nigeria,* Ibadan, Ibadan

University Press.
Poulantzas, Nicos, 1969: 'The Problems of The Capitalist State', *New Left Review*, 58, November/December.
_____, 1973: *Political Power and Social Classes*, London, New Left Books.
POSCO, 1989: Pohangjechu Yishipuyunsha (20 Years History of POSCO), Seoul, POSCO.
Pye, Lucian W., 1985: *Asian Power and Politics: The Cultural Dimensions of Authority*, Cambridge, Harvard University Press.
Radelet, Steven and Sachs Jeffrey, 1997: 'Asia's Bright Future', *Foreign Affairs*, November/December.
Rhee, Jong-Chan, 1994: *The State and Industry in South Korea: The Limits of The Authoritarian State*, London, Routledge.
Rodney, W., 1976: *How Europe Underdeveloped Africa*, Dar-Es-Salam, Tanzania Publishing House.
Rosenberg, N., 1976: *Perspective on Technology*, Cambridge, Cambridge University Press
Rostow, W., 1960: *The Stages of Economic Growth*, Cambridge: Cambridge University Press.
Sakong, I.L., 1993: *Korea in The World Economy*, Washington D.C., Institute for International Economics.
Schatzl, L., 1971: 'The Nigerian Tin Industry', NISER, Monograph Series, No.3.
Smith, Adam, 1966: *An Inquiry into The Nature and Causes of The Wealth of Nations*, New York, Arlington House.
Steinberg, David I., 1995: 'The Republic of Korea: Pluralizing Politics', in Larry Diamond *et. al* eds. *Politics in Developing Countries: Comparing Experiences with Democracy*, Boulder, Lynne Rienner.
_____ 1998: 'Korea: Triumph & Turmoil', Journal of Democracy.
Stewart, F., 1978: *Technology and Underdevelopment*, London, Macmillan (2nd Edition).
Song, Byung-Nak, 1994: *The Rise of The Korean Economy*, Hong-Kong, Oxford University Press.
Teriba, O. and Kayode, M. eds., 1979: *Industrial Development* in Nigeria, Ibadan, Ibadan University Press.
Teriba, O., Edozien, E. and Kayode, M., 1981: *The Structure of Manufacturing Industries in Nigeria*, Ibadan, Ibadan University Press.
The Delta Steel Company, 1986: 'Problems of Delta Steel Company', G.M.'s Welcome address to the Minister of Mines, Power and Steel, Aladja, Warri.
The OECD, 1975: *The Iron and Steel Industry.*
Tsumuri, Y., 1976: *The Japanese Are Coming: A Multinational Interaction of Firms and Politics*, Cambridge, Mass Ballings Publication Company.
Ukwu, I.U., 1988: 'Industrialization and Economic Development in Nigeria: The Significance of SAP' in Structural Adjustment Programme and The Nigerian Economy, Nigerian Economic

Bibliography

Society Conference Proceedings.
UNIDO Reports, 1981: *First Global Study on the Capital Goods Industry: Strategies for Development*, ID/WG, 342/3, Vienna.
──────── 1981: *Feasibility Report on Setting up of a Machine Tool Complex in Nigeria*, DP/NIR.75/001, Vienna.
──────── 1983: *Capital Goods in Perspective: Definition, Importance and Analysis of Factors Affecting Demand*, Working Paper Series No.11.
────────, 1985: *Ad-Hoc Expert Group Meeting on Strategies for More Integrated Development Between Iron and Steel and Capital Goods*, Issue Paper 1, Vienna.
────────, 1988: *Regenerating African Manufacturing Industry: Country Briefs*, Vienna.
Unongo, P., 1980: 'Steel Development and Nigeria's Power Status', NIIA Lecture Series 35.
Usman, B.Y., 1982: 'Behind the Oil Smoke Screen: The Real Causes of the Current Economic Crisis', May Day Speech.
Wallerstein, Immanuel, 1979: *The Capitalist World Economy*, Cambridge, Cambridge University Press.
Warren, Bill, 1973: 'Imperialism and Capitalist Industrialization', *The New Left Review*, 81
Warren, Kenneth, 1975: *World Steel : An Economic Geography*, New York, Crane, Russak and company, Inc.
Weber, Marx, 1964: *The Theory of Social and Economic Organisation*, New York, The Free Press
World Bank, 1987: *Korea: Managing The Industrial Transition*, Vol.II, Washington D.C.
────────, 1993: *The East Asian Miracle: Economic Growth and Public Policy*, New York, Oxford University Press.
Yachir, F., 1988: *The World Steel Industry*, London, Zed.

Table 3.1
Private and Non-Federal Government-Owned Steel Companies in Nigeria: 1998

Company/ Location	Ownership structure	Installed Annual Capacity (crude steel)	Technological Route	Raw Material and source	Product Mix
Private Continental Iron and steel company; Ikeja – Lagos	British (Hong-Kong) / Nigeria	90,000 metric tons	electric arc furnace / rolling	metal scrap-based; billets (local) billets (imported)	Steel, rods, bars, and coils
Federated steel company, Otta	Nigeria /British (Hong - Kong)	25,000 metric tons	electric arc furnace / rolling	metal scrap-based; (local) ;steel billets (imported)	Steel, rods and bars
Universal Steel Company, Ikeja-Lagos	British (Hong-Kong) / Nigeria	90,000 metric tons	electric arc furnace / rolling	metal scrap-based; billets (local) /(imported)	Enamel ware; and steel rods, bars, bin cans
Nigeria-Spanish Engineering Company, Kano	Nigeria /Spanish	100,000 metric tons	Electric arc furnace / rolling	Metal scrap-based; (local) ;steel billets (imported)	Agric-tractors parts; farming tools; steel rods and bars
KEW Metal Industry, Ikeja – Lagos	British (Hong-kong) / Nigeria	20,000 metric tons	electric arc furnace / rolling	mental scrap-based; (local) ;steel billets (imported)	steel rods/bars; enamelwares; bolts/nuts

Company	Ownership	Capacity	Process	Input	Output
Mayor Engineering Company, Ikorodu-Lagos	British (Hong-Kong) / Nigeria	100,000 metric tons	rolling	Steel billets, 100% import-dependent	Steel rods, coils
Kwara Comercial Metal and chemical Company; Ilorin	Nigeria	25,000 metric tons	rolling	Steel billets, import-dependent	Steel, rods and bars
Allied steel Company Ltd. Onitsha	Nigeria	25,000 metric tons	rolling	steel billets, import-dependent	steel, rods/ bars
Mandarin Industry, Lagos	Nigeria /Indian	15,000 metric tons	rolling	steel billets, import-dependent	steel, rods/ bars
Non-Federal Niger steel Company, Emene – Enugu	Anambra State Govt/ German (Ferrostaal)	50,000 metric tons	electric arc furnace / rolling	metal scrap-based; steel billets, import-dependent	steel rods, coils
Qua Steel products Company; Eket	Cross River/ Akwa Ibom State Governments/ Nigeria (private)	100,00 metric tons	rolling	steel billets, import-dependent	steel rods

Sources: Author's Field Works, 1996 and 1998.

Table 3.2
Federal Government-Owned Steel Companies in Nigeria

Company /Location	Annual Installed Capacity (crude steel)	Process	Basic Raw Materials and Sources	Product Mix
Integrated Plants Ajaokuta Steel Projects; Ajaokuta	2.5 million metric tons	electric arc furnace/ rolling	Planned to depend on locally sourced iron-ore, coal and natural gas	steel bar, rods, channels, billets blooms expected to be produced when completed
Delta Steel Company, Ovwian-Aladja, Warri	1 million metric tons	Direct reduction/ rolling	iron ore, limestone, national gas. 100% import dependent on iron ore; gas and limestone sourced locally	steel rods, bass, channels, billets
Rolling Mills Katsina Steel Rolling Coy; Katsina	210,000 metric tons	rolling	billets - to depend 100% on DSC; but import was 90% of its billets	steel rods, coils and bars
Jos Steel Rolling Company, Jos	210,000 metric tons	rolling	billets - to depend 100% on DSC; but 90% import dependent	steel rods, coils and bars
Oshogbo Steel Rolling Company Oshogbo	210,000 metric tons	rolling	billets - to depend 100% on DSC; 90% import dependent	steel rods, coils and bars

Sources: Author's field work to the Companies (Ajaokuta, Ovwian-Aladja, Katsina, Jos and Oshogbo) in 1996.

Table 3.3

Goverment-Owned Steel Company in South Korea

Company /Location	Annual Installed Capacity (combine) :(crude steel)	Process	Basic Raw Materials	Product Mix
Integrated Steel Plants Pohang Iron and Steel Coy. Ltd; Site1:Koedong-dong, Pohang,Kyungsang; Site 2: Kwangyang	38,903,000 metric tons	conventional blast furnace, COREX; and mini mill	iron ore, coal and limestone. 100% import dependent on iron ore and coal; locally sourced limestone	Wire rods, hot and cold rolled coils and sheets, pickled and oiled coils, thin plate and coils; electric steel sheets and strips. Eletrogalv. Coils, sheets, stainless rolled coils and sheets,coated organic steel sheets, etc.

Sources: Author's field trip to POSCO-Pohang, June 3, 1997 Kwangyang, June 20, 1997

Table 3.4

Major Private Steel Companies in South Korea

Company/Location	Ownership structure	Installed Annual Capacity (crude steel)	Technological Route	Raw Material and source	Product Mix
Hanbo Steel Company Ltd; Daechi-dong, Kangnam-Ka, Seoul	Korean	4,000,000 metric tons	Electric arc furnace/ direct reduction/ COREX / rolling	Iron ore, coal and pig iron; import dependent	reinforcing bars, round bars, angles, hot rolled coils and sheets
Inchon Iron and Steel Coy. Ltd; Songyum-dong-Ku; Seoul	Korean	3,630,000 metric tons	electric arc furnace/ rolling	scrap-based; (local and import sources); billets and steel sheets depend on POSCO	rail road rails; H-beams; sheet pipes, Angles, Channels, stainless steel; steel casting
Dongkuk-Steel Mill Coy. Ltd; Suha-dong, Chung-Ku; Seoul	Korean	2,500,000 metric tons	electric arc furnace/ rolling	scrap-based; (local and import sources); billets/sheets; largely dependent on POSCO; and minimal importation	angles, channels, round bars, flat bars, steel plates, general structure and ship building
Kangwon Industries Limited; Shinmoon-ro 2-Ka, Chongno-Ku, Seoul	Korean	2,000,000 metric tons	electric arc furnace/ rolling	scrap-based; (local and external sources); billets/sheets; largely dependent on POSCO; and minimum imports	Angles, channels, I-Beams, H-Beams, rails, rail accessories, and track shoes, flat bars
Hankook Steel & Mill Company Ltd; Yuhyun-ri, Gunbook-myun, Haman-Gun, Kyongsangnam-do	Korean	1,580,000 metric tons	rolling	billets; depend largely on POSCO	reinforcing bars

Sammi Steel Coy Ltd; Daechi-3 dong, Kangnam-Ku, Seoul	Korean	800,000 metric tons	electric arc furnace/ rolling	metal scraps; steel(local) billets/sheets (POSCO)	stainless steel sheets, special steel bars, section rods, steel casting/ forgings
Dae Han Steel Mill Coy. Ltd; Shinlyung-dong, Saha-Ku, Pusan	Korean	600,000 metric tons	electric arc furnace/ rolling	metal scraps(local /foreign); steel billets/sheets, depend on POSCO	reinforcing bars, round bars, cold finished steel bars
Kia Steel Coy. Ltd; Soryong-dong, Kusan Cheollabuko	Korean	520,000 metric tons	electric arc furnace/ rolling	metal scraps(local /foreign); steel billets/sheets, depend on POSCO	special steel (rolled and forged) bars, crame, wheel, auto-parts, (crank shafts axle)

Sources: Author's field work in South Korea (POSRI, Seoul; Inchom; Pusan;Kosa, Seoul)

* Hanbo steel company became bankrupt in March 1997.

Table 4.1 DSC: Product Mix and Status of Production

Nature of Product	Specification (mm)	Status
Rounds	12 - 50	Produced at 6% of installed capacity, shut down indefinitely since 1996
Squares	12 – 50	The line collapsed during trial production and was never repaired
Plats	20 – 60	It was only thought of, but was never installed
Tees	20 – 60	Production line crashed during trial run, and thereafter shut down
Equal Angles	20 x 3 – 65 x 11	Unit was never installed
Equal Angles	65 x 5 x 9	Unit was never installed
Channels	40 x 20 – 100 x 50	Unit was never installed
1 – Beams	80 and 180	Unit was never installed

Source: Author's several field trips to the Company in 1987, 1990, 1995 and 1997

Table 4.2   Position of Raw Materials for Nigerian Steel Industry

Ore	Requirements of the Nigerian Steel Plants			Locally Available Reserve	Projected Life Span	Remarks
	Ajaokuta Steel Plant	Delta Steel Plant	Total Requirements			
1	2	3	4	5	6	7
Iron Ore	2 Million tonnes (Fe 63%) p.a.	1.5 Million tonnes (Fe 67%) p.a.	3.55 million tonnes p.a.	Ferruginous Quartzite Fe. 38% Itakpe - 300m. Tonnes Ajabonoko Hill -- 50m. " Chokochoko – 12m " Agbado Okudu – 24m. " Tajimi - 20m. " Anomaly K3 - 30m. " 436m tonnes	23 years	Based on generalised 50% confidence level
Cooking Coal	1.2 million tonnes (V.M. – 18.32% Ash 9% Sulphur 13.0%	5,000 tonnes	1.7 million tonnes	Lafia - 162 m. tonnes	19 Years	Based on 32m tonnes cokable coal. Lafia coal field steam 12 – 50% confidence level.
Refractory Clay	70,000 tonnes Al_2O_3 – 37.39% FeO_3 – 2.5%	50,000 tonnes	120,000 tonnes	Onibode – 2.8m tonnes	10 Years	Based on External interest
Dolomite	255,000 tonnes MgO – 19% SiO_2 - 1.0%	-	255,000 tonnes	Burum - 4m tonnes	5 Years	Other industrial usage
Limestone	650,000 tonnes CaO-535% SiO_2 – 1%	130,000 tonnes $CaO – 535\%$ SiO_2 – 1%	780,000 tonnes	Jakura - 33m tonnes Mfamosing – 26m. " Ubo - 20m tonnes 78m. tonnes	33 Years	Based on competitions from other industries.
Manganese	90,000 tonnes	-	90,000 tonnes	Mallam Ayuba: tonnage not yet determined	-	Exploration in progress
Bauxite	13,000 tonnes Al_2O_3 – 40%	-	13,000 tonnes	Oju: tonnage not yet determined	-	Exploration in progress

Source: Steel Raw Material Exploration Agency, Kaduna, 1987, 1992, 1997.

Table 4.3
POSCO: Import of Iron Ore and Coking coal by Volume and Country

Steel Raw Material	Country	Volume (tons) 1995
Iron Ore	Australia	17,732,000
	Brazil	8,335,000
	India	3,081,000
	Peru	1,098,000
	Others	2,938,000
	Sub Total	33,184,000
Coking Coals	Australia	8,177,000
	Canada	3,782,000
	USA	1,967,000
	Others	1,896,000
	Sub Total	15,822,000

Total Volume of Imported Steel Raw Materials = 56,533,000
% of Iron Ore of the Total = 59
% of Coking Coal of the Total = 28

Source: POSCO.

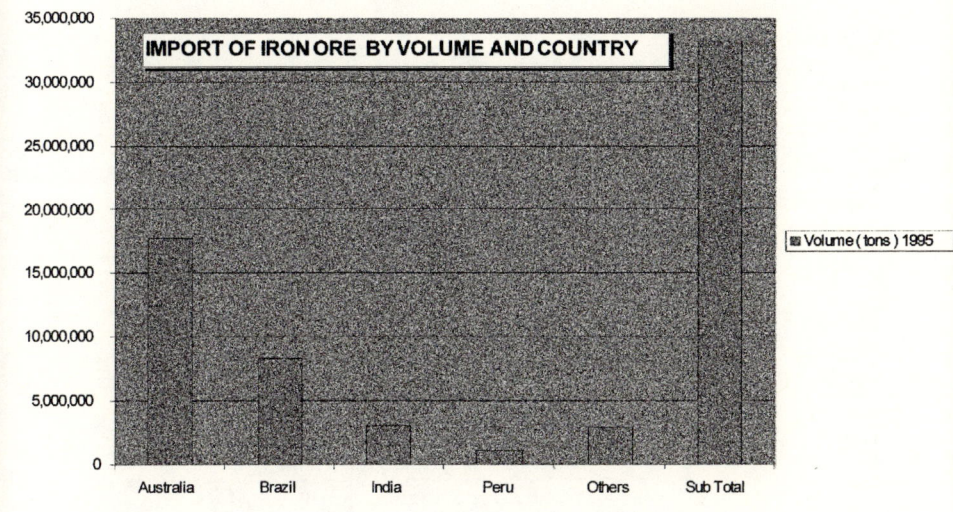

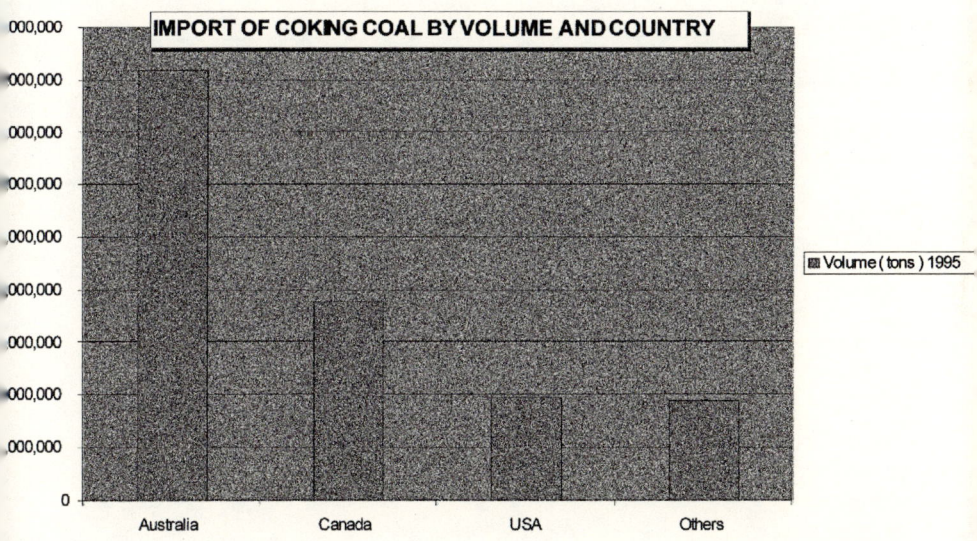

Table 4.4
POSCO: Units, Facilities and Installed Capacities, 1995

Units	Facilities	Capacities * (million tons)
Iron and Steel Making	Iron Making	21.74
	Steel Making	40,19
Rolling	Section	4.32
	Reinforced Bar	9.75
	Bar, Wire and Rod	3.86
	Plate	3.55
	Hot Rolled Sheet	19
	Cold Rolled Sheet	7.81
	Electric Sheet	0.26
	Stainless Steel, Hot Rolled Sheet	0.68
Others	Galvanized	3.10
	Tin Plate	0.72
	Pipe and Tube	3.90

Source: Author's field trip to POSCO, Pohang; POSRI, Seoul, May/June 1997.

*Note   Although projections on the expected increases into the year 2000 were made in 1995, no expansion had taken place by 1997. It was only in the production of reinforced steel bars that a decrease was projected.

Figure 4.1 Flow Chart of Iron and Steel Making: Blast Furnace

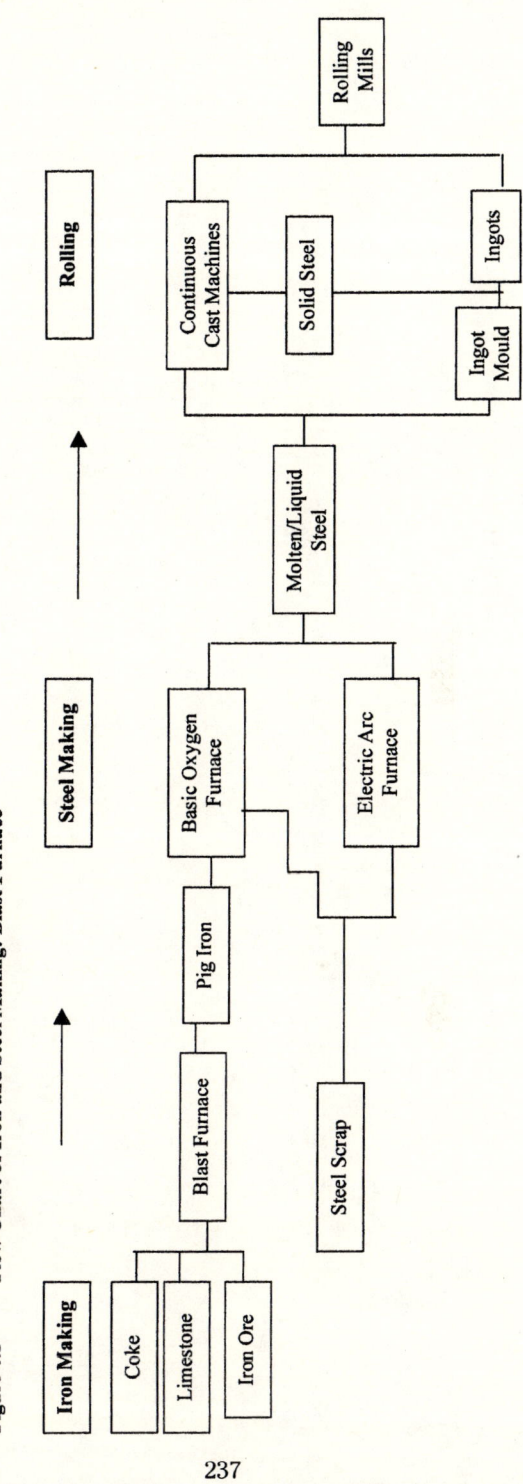

237

Figure 4.2 Production Chart of Iron and Steel Making: Direct Reduction Process.

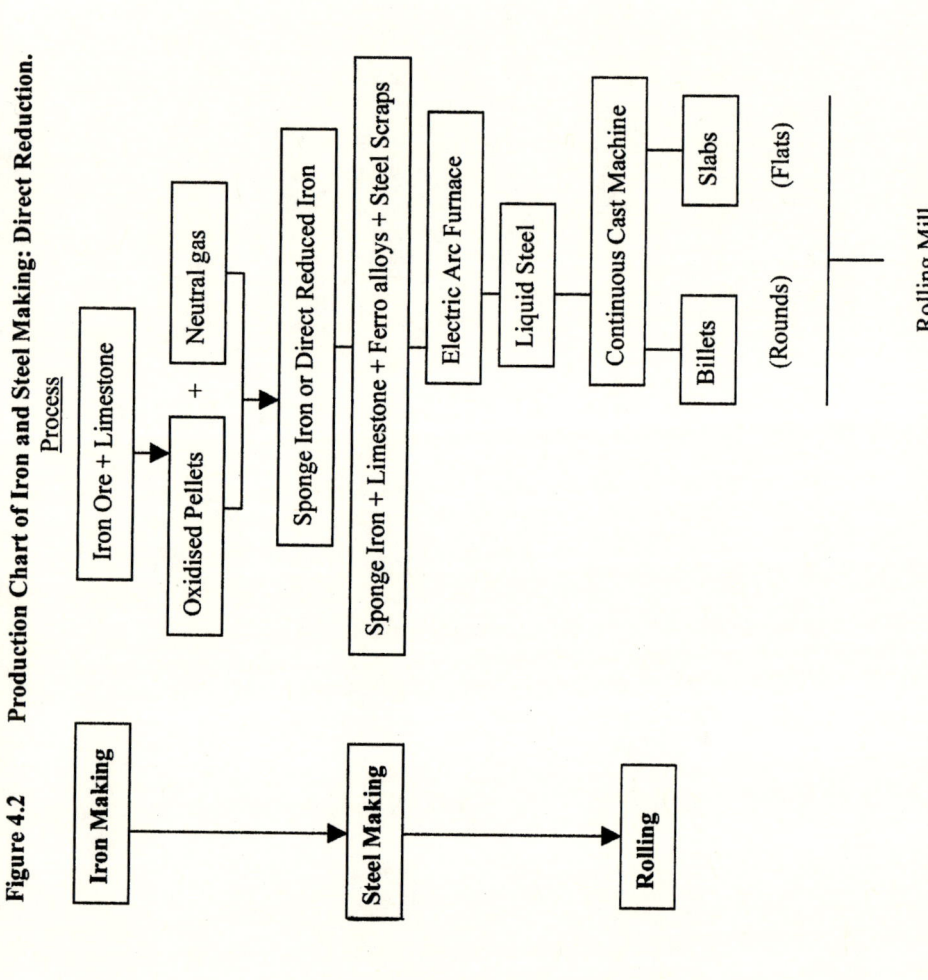

Figure 4.3   Flow Chart of Iron and Steel Making: Mini-Mill.

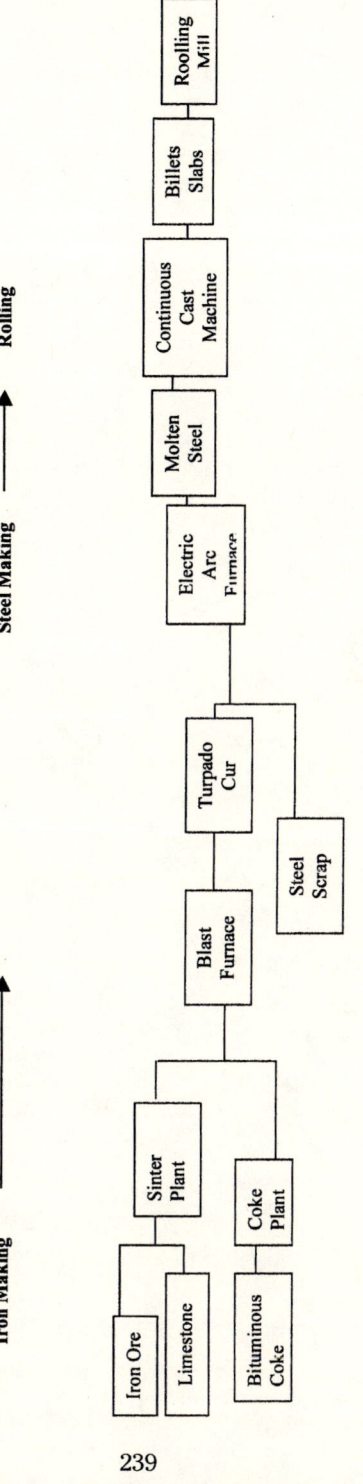

Figure 4.4   COREX Process of Iron and Steel Making

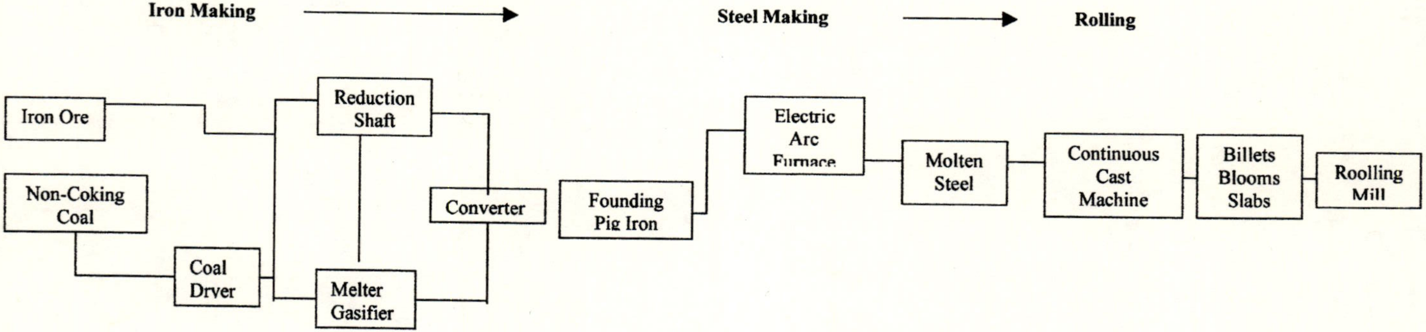

Table 5.1
Steel Self-Sufficiency Ratio (SSR) of some Selected Steel-Producing Asian / African Countries in Percentage, 1996

Countries	SSR (%)
Japan	121
China	92
South Korea	99
India	78
Taiwan	48
Singapore	10
Nigeria	5
Egypt	15
Zimbabwe	5

Sources: ISSI, Steel statistical Year Book, (Brussels, 1996); AISA, Reports on African Steel Industry, (Abuja, 1996).

Table 5.2

Per Head Crude Steel Consumption by Country (Selected ASEAN and African Countries (1995)

Unit in Kg/ head	Country
0 –20	Vietnam, Ghana, Kenya
21 –30	India, Nigeria, Indonesia, Zimbabwe
31 –100	China, Philippines
101 –200	Thailand
201 –500	Malaysia, South Africa
501 – 1000	Japan, South Korea
1001 +	Singapore, Taiwan

Source: Abstracted from the Steel Statistical Year Book, IISI, 1995

Table 5.3

Total Production of Intermediate Steel Products by Public Steel Companies 1986 - 1995
Product Profile (Quantity in metric tonnes)

Year	Rods and bars for Reinforcement	Wire Rods in Coils	Other rolled Products
1986	31,314	32,371	66,944
1987	9,826	13,908	53,762
1988	6,492	5,445	60,264
1989	n.a	n.a	49,030
1990	n.a	n.a	55,534
1991	n.a	n.a	n.a
1992	1,500	500	n.a
1993	n.a	n.a	n.a
1994	300	n.a	n.a
1995	n.a	n.a	n.a

Source: Field work, July - October 1994; 1996.
Delta Steel Company, Aladja-Warri
Oshogbo, Jos and Kastina Steel Rolling Companies,
Federal Office of Statistics, Lagos.
n.a = not available

Table 5.4

Structure of Intermediate and Capital Iron and Steel Products/ Goods in Nigeria 1982 – 1991: Imports Year (Quantity in Metric Tonnes)

Commodities	1982	1983	1984	1985	1986	1987	1988	1989	1990	1991
Long Products (Total)	**1,779,000**	**276,000**	**357,000**	**309,000**	**82,000**	**79,000**	**108,000**	**63,000**	**50,000**	**79,000**
Ingots and semi] Railway Track Material	24,000	14,000	19,000	227,000	54,000	33,000	30,000	20,000	4,000	9,000
Heavy Sections	3,000	1,000	nil	nil	nil	1,000	5,000	3,000	4,000	1,000
Bars & Light Sections	351,000	85,000	58,000	14,000	15,000	6,000	6,000	3,000	6,000	10,000
Concrete Reinforcing Bars	226,000	57,000	26,000	10,000	3,000	5,000	15,000	10,000	5,000	6,000
Wire Rod	869,000	67,000	25,000	38,000	7,000	25,000	46,000	21,000	22,000	37,000
Drawn Wire	136,000	32,000	35,000	4,000	1,000	1,000	3,000	4,000	7,000	14,000
	70,000	20,000	194,000	16,000	2,000	8,000	3,000	2,000	2,000	2,000
Flat Products (Total)	**1,298,000**	**1,038,000**	**729,000**	**224,000**	**95,000**	**173,000**	**63,000**	**141,000**	**162,000**	**225,000**
Plates less than 3mm	580,000	292,000	125,000	36,000	21,000	15,000	15,000	6,000	12,000	14,000
Sheets less than 3mm	336,000	32,000	28,000	123,000	41,000	114,000	22,000	103,000	124,000	172,000
Strip	38,000	7,000	3,000	4,000	8,000	2,000	3,000	8,000	1,000	1,000
Tin plate & Blackplate	150,000	24,000	21,000	45,000	12,000	21,000	10,000	21,000	20,000	30,000
Galvanized sheets	194,000	683,000	552,000	9,000	10,000	18,000	8,000	3,000	4,000	5,000
Other coated sheet	n.a	n.a	n.a	7,000	3,000	3,000	5,000	1,000	1,000	3,000
steel Tubes (Total)	**347,000**	**1,038,000**	**729,000**	**224,000**	**95,000**	**173,000**	**63,000**	**141,000**	**162,000**	**225,000**

Sources: Extracted from **Annual Abstract of Statistics (Lagos: FOS) 1996**; and author's field work: Delta Steel Company; Selected Private Steel Companies; Oshogbo, Jos and Kastina Steel Companies. 1996

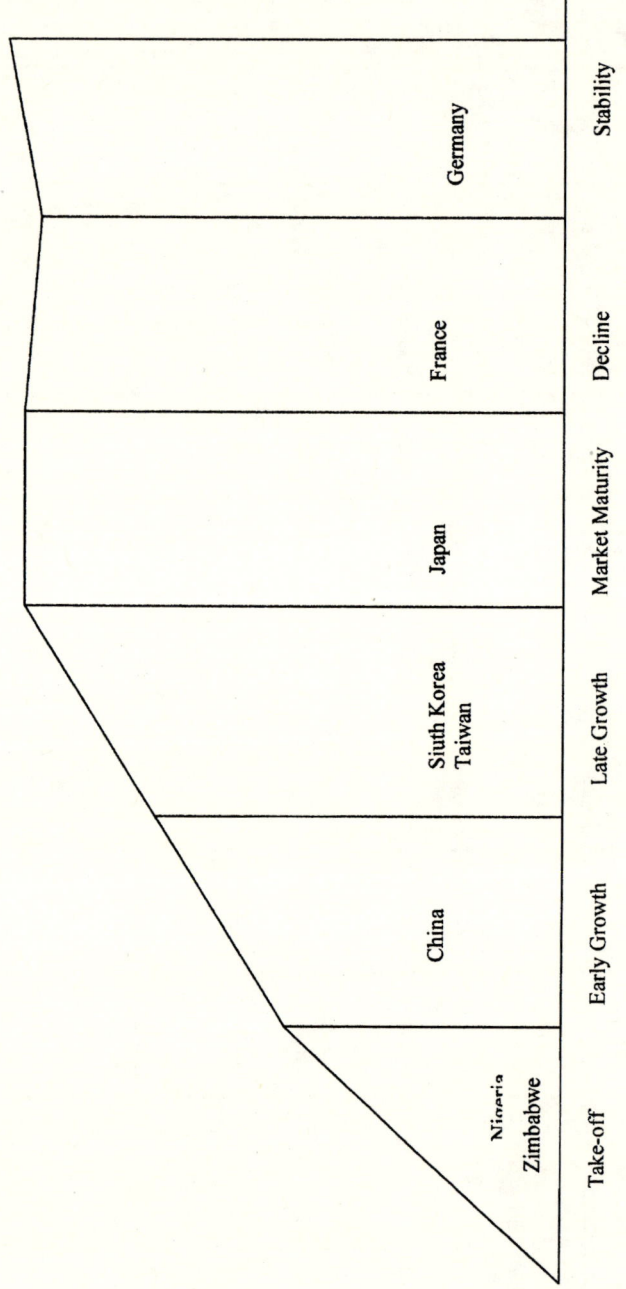

Table 5.5

Light Industries;
Metal Manufactures and Fabricators using Intermediate and Capital Goods and their products

Company	Address/Ownership	Sectoral Group	Sub-Sector	Nature of Basic Raw Materials and Sources	Products manufactured/Broad names
Nigeria Foundries ltd.	Ilupeju, Lagos; private (1989)	Basic metal	Foundry Engineering	Light steel sheets, bars, beams and rods; import dependents	Hand operated water pumps, pressure pipes for water/waste drainages, valves and other industrial equipments
Road Engineering & Foundry Limited	Ilupeju, Lagos; private (1952)	Basic metal	Foundry Engineering	Recycling ferrous and non-ferrous scraps; import dependent.	Casting both ferrous and non-ferrous metals; based on samples and specifications
Armeco-Arewa Metal Construction Limited	Kaduna public/private (ratio not available)	Basic metal	Contruction engineering/metal manufacturing	Light/medium steel flats; import dependent	Agric silos; soft beer canners; bodies od tankers and trailers; surface & underground storage tanks.
Anchor Products Limited	Ikeja, Lagos; private; (1978)	Basic metal	Contruction engineering/metal manufacturing	Special steel, bars and rods, ferrous and non-ferrous alloys; and their scraps; import dependent	Molds for plastic and aluminium industries; machine parts such as shafts, bearings and gears; fabricates simple machines and equipment according to specifications

Agricultural Machinery Manufacturing Co. Ltd.	Ketu, Lagos; private; (1984)	Basic metal	Engineering and Fabrication	Light/medium steel flats; import dependent	Spades, shovels, matchets, rakes and travel; brand name is **bull**
BACO Engineering Coy. Limited	Aba, private; (1972)	Basic metal	Fabrication	Light/medium steel flats; import dependent	Underground tanks, door hinges, shoe shanks, bed hooks; brand name is **BACO**
Hoesech Pipe Mills Nig. Limited.	Ikeja, Lagos; private; 40% Nigeria; 60% foreign (1974)	Basic metal/engineering	Fabrication	Light steel flats (galvanized); import dependent	Galvanized steel buckets, storage/waste bins, metal mesh; brand name is HOESCH
Jos Hansen & Soechne Nig. Limited	Ojota, Lagos private: - Nigeria 60%; foreign 40% (1955)	Basic metal/engineering	Fabrication	Light and medium steel flats, non-ferrous metals, cables; import dependent.	Sale of imported centrifugal Pumps, hand operated pumps, and fabrication of their parts and rewinding of electric motors: brand name is KSB
Kew Metalworks Ltd.	Ikeja, Lagos; private; (1976)	Basic metal	Fabrication	Special light and medium steel bars, rods and flats; import dependent	Steel window/door frames; construtional products such as bars - J,Z,T and angles.

		Basic metal/engineering	Fabrication/engineering foundry		
Light Machines Ind. Nig. Limited	Ikeja, Lagos, private (1971)			Light and medium steel flats, non-ferrous metals and alloys; import dependent.	Steel metals for forming bodies: spare parts for machines and moulds.
Metal Furniture Nigeria Limited	Ikeja, Lagos; private (1962)	Basic metal	Fabrication	Special light and medium steel bars, rods and flats; import dependent	Steel office equipments
Modern Engineering Works Ltd	Sango Otta, private; (1977)	Basic metal/Engineering	Fabrication	Ferrous and non ferrous metals, special steel flats; import dependent.	Molds for plastic industries
Nigeria Gas Cylinders Co. Ltd.	Agodi, Ibadan; private; (1977)	Basic metal	Fabrication	Special flat steel; import dependent	Gas cylinder for both indus Trial and domestic uses.
Pressed Metal Works Co. Ltd.	Ijora, Lagos; private -40% Nigeria; 60% (1960)	Basic metal	Fabrication	Light and medium steel flats; import dependent.	Vehicle bodies (pmv comm Ercial);storage tanks (pmv structures)and agricultural implements (pmv agricultural)
SCOAIARD (A Division of SCOA Nig.Limited)	Lagos; private and public; 60% Nigeria; 40% foreign	Basic metal	Fabrication	Special steel, light and medium steel flats; import dependent	Bodies for cars vehicles and refuse bins
SETEMEC Limited	Isolo, Lagos; private- 40% Nigeria; 60% foreign	Basic metal	Fabrication	Special light and medium steel flats; import dependent.	Gas cylinder and LPG tanks
Steel Works Limited	Oluyole, Ibadan; private;73% Nigeria; 27% foreign (1975)	Basic metal	Fabrication	Light and medium steel aluminium flats ;import dependent	Aluminium roofing, cladding and construction steel products.

Company	Location, ownership; (year)	Sector	Sub-sector	Product type	Finished products
Baltic Engineering Group Limited	Benin City, private-100% Nigerian;(1975)	Basic metal	Engineering / fabrication	Light/medium steel and aluminimum flats and rods; import dependent	Bakery equipment. Hospital equipment, mechanical parts; brand name is BALTIC
Crittal-Hope Nig. Ltd	Mushin, Lagos; private; 50% Nigeria; 50% foreign; (1958)	Basic metal	Engineering / fabrication	Light/medium steel and aluminimum flats and rods; import dependent	Steel and aluminium door and window frames; louvre channels
Christer Engineering Works Nig. Ltd	Oshodi, Lagos; private; (1980)	Basic metal	Fabrication	Light/medium steel flats;import dependent	Silencer and mufflers for vehicles, generators and engines.
Crocodile Matchets Nigeria Limited	Trans Amadi,Port Harcourt; private;40% Nigeria; 60%foreign; (1965)	Basic metal	Fabrication	Light steel flats; import dependent	Cutlass; brand name is CROCODILE
Ekene Dili Chukwu (Steel Structures) Ltd.	Awada, Onitsha; private; (1976)	Basic metal	Engineering/ fabrication	Light and medium steel flats; import dependent.	Kerosene and petrol tanks; bodies for tippers, trailers, refuse and loaders
Eldlleoseka Investment Co. Ltd	Nkpor, Onitsha; private; (1983)	Basic metal	Fabrication	Light/medium steel and aluminium flats;rods and coils; local and import dependent.	Wheel barrow, wirefence and rivet nails

Sources:

Author's Field Work: July August And September 1994; September 1996;and December 1997.

Universal Steel Company, Ikeja - Lagos.
Maunfacturers Association of Nigeria, Ikeja-Lagos.
The Various Industrial Areas of Lagos
The Industrial Areas of Otta, Ogun State
Nigerian Chamber of Commerce, Agriculture, Mines, Maryland, Lagos.
Federal Office of Statistics, Lagos.

Table 5.6

Heavy Industries:

Motor Vehicle Assembly and other companies using products of and fabricating goods for the Heavy Industries

Company	Address/ Ownership	Sectoral Group	Sub-Sector	Nature of Basic Raw Materials and Sources	Products manufactured/Broad names
Brossette Manufacturing Co. Ltd.	Kakuri, Kaduna; private (1981)	Automotive manufacturing	Fabrication of automotive parts	Flat steel sheel, bars and rods; import dependent.	Fuel tanks and pedal systems.
Cento International Co. Ltd.	Nkwo Akwu, Nnewi; Private, 100% Nigeria (1982)	Automotive Manufacturing	Fabrication of automotive parts	Aluminium coils, rods and bars; import dependent.	Front grills for peugeot **404,504** and **505**; brand name **CENTOCO AUTO.**
Coksee Engineering Works Limited	Alakuko, Lagos; Private 100% Nigeria (1976)	Automotive Manufaturing	Fabrication of auto parts	Special flat steel; import dependent.	Vehicle bodies for trailers and tankers.
Fabinna Nigeria Ltd.	Port Harcourt Road Aba Private;100% Nigeria.(1978)	Automotive Manufaturing	automotive parts	Cast iron and alloys; flat steel and bars; import dependent; recycling scraps of brake shoes and pad lacally.	Brake shoes, pads and linings; brand name - **FABINNA**.
Ferodo Nigeria Ltd.	Oluyole, Ibadan, private, 40% Nigeria; 60% foreign (1976)	Automotive manufaturing	auto parts	Cast iron and alloys; reconditioning local scraps of brake pads and lining; also import dependent.	Brake lining and disc pads.
Fichitel & Sachs (W.A) Ltd.	Apakun,Oshodi - Lagos;private; Nigeria 44%; foreign 56% (1982)	Automotive manufaturing		Cast iron and alloys; bars and flat steel; import dependent	Clutches for passenger cars, lorries, buses and agric tractors; brand name; **SACHI CLUTCHES.**

Company	Location/Ownership	Sector	Sub-sector	Inputs	Products
International Parts Industrial Ltd	Kachia, Kaduna; private;40% Nigeria;60% foreign (1982)	Automotive manufaturing	auto components	Cast iron and alloys; aluminium coils and rods;special steel flats and bars; import dependent.	Clutches (cover and plates) for cars: alternators.
Nivafer Steel Construction Co. Ltd	Agege, Lagos; private,60% Nigeria;40% foreign	Automotive manufaturing	auto components fabrication	Flat steel sheet (special); import dependent.	Bodies for trailers and lorries.
Radiators Nig. Ltd	Trans Amadi, Port Harcourt; private; 95% Nigeria; 5% foreign(1979)	Automotive manufaturing	auto components	Cast iron and alloys; aluminium coil and rods; import dependent; reconditioning scraps locally	Radiators for peugeot 504 and 505
Comrade Cycle Co. Nigeria Ltd.	Industrial Area, Zaria; private;70% Nigeria;30% foreign (1979)	Bicycles manufacturing	Bicycles manufacturing	Complete knocked down parts (ckd); import dependent	Bicycles
Raleigh Industries Nig. Ltd.	Bompai, Kano; private ;60% Nigerian; 40% foreign (1974)	Bicycles manufacturing	Bicycles manufacturing	Complete knocked down parts (ckd); import dependent	Tricycles and bicycles; brand names **RALEIGH ROBIN HOOD** and **RUDGE** for bicycles.
Boulous Enterprises Limited	Acne,Ogba-Ikeja; Lagos private; 70% Nigeria; 30% foreign (1964)	Motor Vehicles manufacturing	Motorcycles manufacturing	Complete knocked down parts (ckd); import dependent	Motor cycle; brand name is **SUZUKI**
Yamaha Nigeria Limited	Abule-Egba, Agege-Lagos; private; 60% foreign; 40% Nigeria (1980)	Automobile	Motorcycles manufacturing	Complete knocked down parts ; import dependent	Motor cycle; brand name is **YAMAHA**
Anambra Motor Manufaturing Co. Ltd	Emene-Enugu; 60% Nigeria; 40% foreign; (1975)	Automobile assembly	Vehicle assembly	Complete knocked down parts ;flat steel; import dependent	Bodies for mercedes benz commercial vehicles.

Federated Motor Industries	Apapa, Lagos; 60% Nigeria; 40% foreign (1959)	Automobile assembly	Vehicle assembly	Complete knocked down parts ;flat steel; import dependent	Bodies for Bedford vehicles.
SCOASSEMBLY (A Division of SCOA)	Kaduna; public /private (1972)	Automobile assembly	Vehicle assembly	Complete knocked down parts ; import dependent	Peugeot cars and vehicle brand names are **PEUGEOT 504, 505** etc.
Volkswagen of Nigeria Ltd.	Ojo-Lagos;public /private. 51% foreign; 49% Nigerian(1973)	Motor Vehicles Assembly	Motor Vehicles Assembly	Complete knocked down parts ; import dependent	Passenger cars; light commercial vehicles; **SANTANA JETTA** etc
SCOASSEMBLY (A Division of SCOA)	Apapa, Lagos; public/private; 60% Nigeria; 40% foreign (1950)	Motor Vehicles Assembly	Motor Vehicles Assembly	Complete knocked down parts ; import dependent	Light commercial vehicles (pickups) brand name is **PEUGEOT.**
Steyr Nigeria Limited	Bauchi, public/ private;60% Nigeria;40% foreign (1976)	Motor Vehicles Assembly	Motor Vehicles Assembly	Complete knocked down parts ; import dependent	Heavy and light commercial vehicles; agric tractors; brand name is STEYR

Sources: Manufacturers Association of Nigeria: Lagos, Kano, Enugu and the companies at Lagos,Ibadan,Otta and Kaduna July-October, 1994,1996

Author's Field Work:

Table 5.7

Structure of Crude steel Production and Consumption in South Korea (1994-1996) in metric tons

Production/ Consumption	Year			% Annual Growth change
	1994	1995	1996	
Total Domestic Demand	42,160,000	46,862,000	50,210,000	7.1
Local Consumption	32,188,000	37,306,000	40,220,000	7.5
Domestic Production	33,745,000	36,772,000	39,030,000	6.1
By process / technological route:				
*BOF	21,610,000	22,873,000	23,500,000	2.7
**EAF	12,135,000	13,899,000	15,530,000	11.7
Export	9,972,000	9,556,000	9,990,000	4.5
Import	8,415,000	10,180,000	11,180,000	10.8

Source: Based on Author's field work at, POSCO, Private steel compasion, POSRI, and KOSA in Seoul, March-August, 1997;

Note: *BOF = Blast Furnace
  **EAF = Electric Arc Furnace

Table 5.8

Steel Production and Consumption (finished steel products) in South Korea (1994-1996) in tons

Production/Consumption	Year			% Annual Growth change
	1994	1995	1996	
Total Local Demand	39,783,000	44,210,000	47,320,000	7.0
Domestic Consumption	30,510,000	35,529,000	38,200,000	7.5
Total Local Production	36,137,000	39,258,000	42,550,000	8.4
Export	9,273,000	8,681,000	9,120,000	5.1
Import	3,646,000	4,952,000	4,777,000	8.4

Sources: Based on Author's field work at POSRI and, Kossa Seoul, June, 1997.

Note: Total Local steel Demand = Domestic Consumption + Export
Or
Total Production + Import

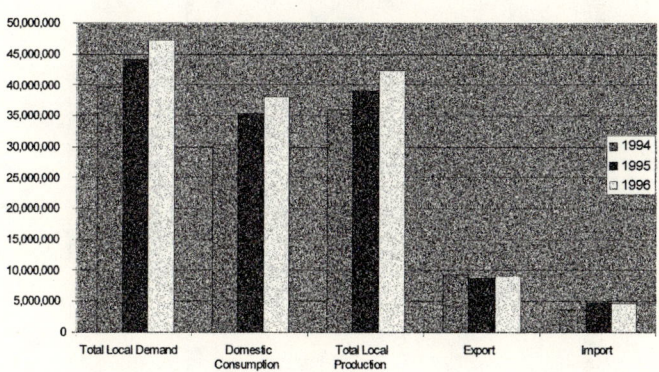

STEEL PRODUCTION AND CONSUMPTION IN SOUTH KOREA (1994-1996)

Table 5.9

Nigeria and South Korea:

Comparism of steel consumption by product and industries

Product	Nigeria Year ('000 tons)		South Korea Year ('000 tons)			
	1980 - 1985	1990 – 1995	1980	1985	1990	1995
Long	200	100	2,589	4,748	10,034	17,343
Flat	nil	Nil	2,424	5,087	9,664	17,642
Fabricated Metals	20	8	101	186	356	544
Total	220	108	5,114	10,021	20,054	35,529
	% of sectorial consumption		% of sectorial consumption			
Industry	1980-1985	1995	1980			1995
Construction	95	96	57.1			47.1
Manufacturing:						
Auto	3	2	5.3			12.3
Machinery	0.5	0.3	4.4			8.3
Electrical Equipment		0.4	3.8			7.1
Ship-building			4.9			5.8
Fabrication	1.3	1	23.6			15.5
Others	0.7	0.3	0.9			3.9
Total	100	100	100			100

Source: Author's field work to FOS, NACCEMA, MAN, Lagos; AISA, Abuja Nigeria 1996; POSRI, KOSA, Seoul, Korea, 1997

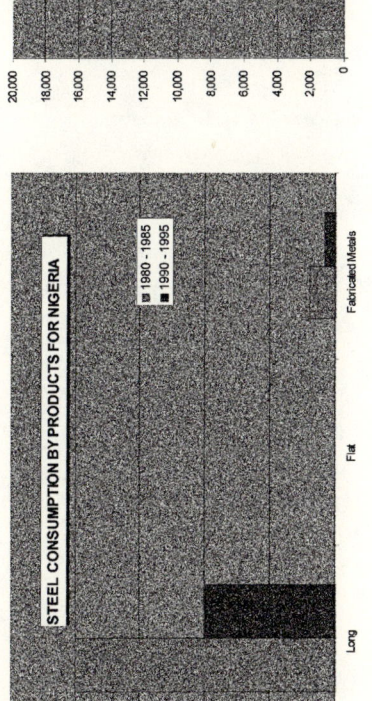
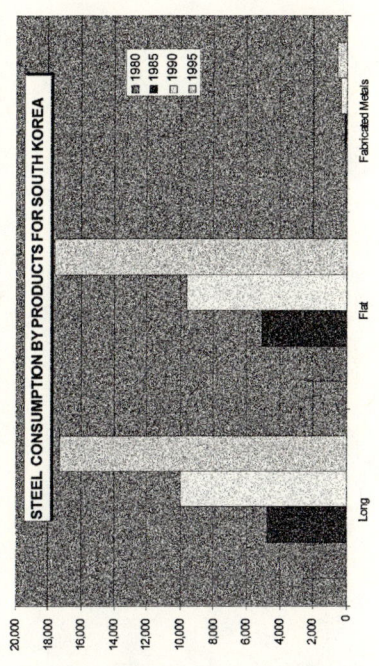
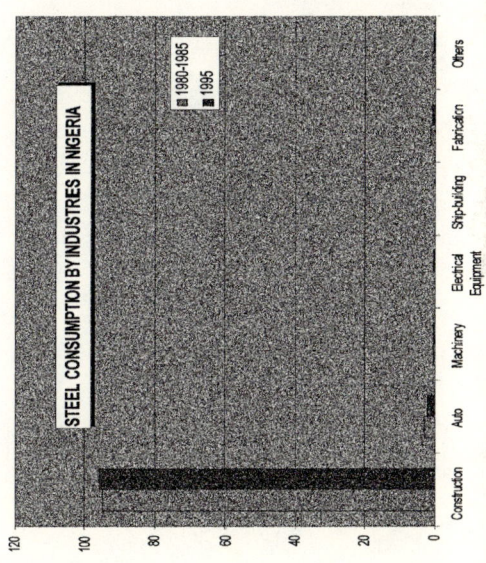
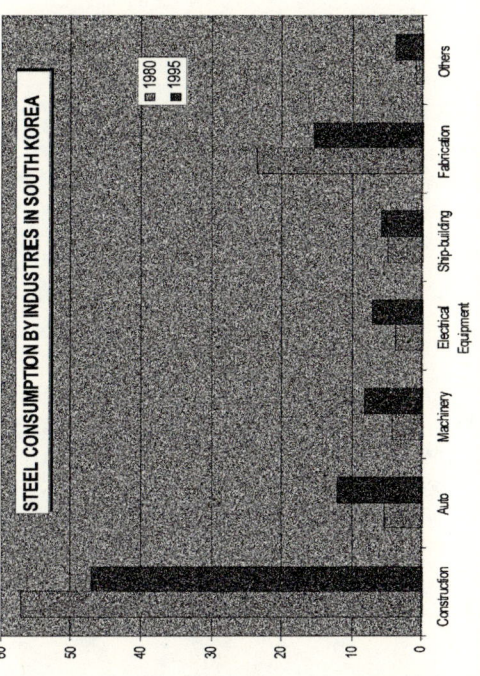

255

Table 5.10

Steel Consumption by Industry in Japan, 1994 (%)

Industry	% Share
Contrution	43.3
Manufacturing:	
Auto	23.3
Ship-building	4.9
Machinery	10.1
Electric Equipment	6.1
Others	13.3
Total	**100.00**

Source: Abstract from steel statistical year book, 1995

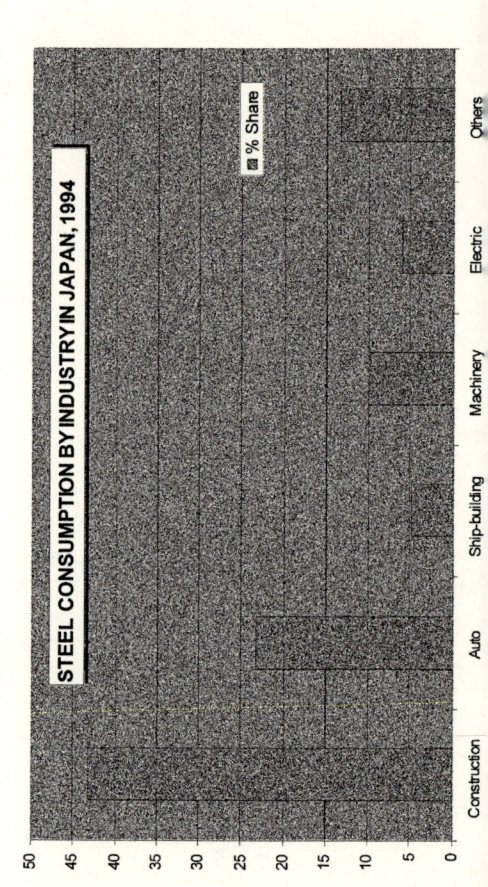

Table 5.11

POSCO's Major Subsidiaries by Sector, Sub-sectoral Activity and product, 1997

Company	Sectoral Group	Sub-Sectoral Activity	Major Products Intermediate and Capital goods)
POSCO Machi-ery and Engine-ering Co., Ltd. (POSMEC), estalished in 1982	Steel manufacturing and Construction	Design, manufacturing, instal and maintain steel making machineries	Steel-making machines and conveyor steel rollers
POSCO Maintenance Co., Ltd. (POS-M) established in 1987	Steel, manufacturing and construction	Manufacture maintenance parts, and steel construction	Parts of the blast furnance and other steel making machines
POSCO Refractories Co., Ltd. (POSREC) established in 1971	Steel, manufacturing and construction	Manufacture, construct and repair furnaces base refractory materials	Part of all furnances, and refractory materials
POSCO Engineering & Construction Co. Ltd established in 1982 (POSEC)	Engineering and construction	Building steel works, steel bridges, ports, skyscrappers	Parts of blast furnances, direct reduction plants and mini-mills
POSCO Energy (POS-Energy)	Enery, manufacturing and construction	Design and produce parts of, and install and operate thermo- electric power plants, and energy resources development	Parts of thermo-electric power plants

Company/ Location	Ownership/Equity	Sectoral Group	Activity	Major product
Pohang Steel America Crop (POSAM), headquaters, New Jersey, USA (1984)	100% POSCO equity : US $ 246.8m	Services (Scale)	Trade in steel products, and operate POSCO's subsidiaries / affiliates in the USA	-
USS-POSCO Industries (UPI); HQ. Pittsburg, California, USA (1986)	Joint Venture: POSAM 50% United States steel 50% Equity: US $ 388.0m	Manufacturing and service (scale)	Manufacturing and sales of cold-rolled steel products	Cold-rolled steel products'
POSVINA Co., Ltd (POSVINA), HQ. HO Chi Minh, Vietnam (1992)	Joint Venture : POSCO 50%; SSC F Vietnam 50% equity US $ 3.9 million	Manufacturing and service (scale)	Manufacturing and sales of galvanized steel products	galvanized steel products
VSC- OPSCO Steel Corp.(VPS), HQ. Haiphong, Vietnam (1994)	Joint Venture : POSCO 35%; VSC 34% Haiphong Engineering & Construction Materials corp 10%, POSTRADE 5% HQ. Haiphong, Vietnam Equity US $ 16.8 Millon	Manufacturing and service (scale)	Manufacturing and sales of bar /wier rods	Steel bar / wire

Company/ Location	Ownership/Equity	Sectoral Group	Activity	Major product
POCSO Asia Company Ltd. (PQA) HQ.. Hong-Kong, (1985)	Joint Venture : POSTRADE 60% DOOSAN 40% Equity : hk $ 6.2 million	Service (Export)	Exporting finished steel products to China and South East Asia	-
Pohang Steel Australia PTY. Ltd (POSA) HQ. Sydney, Australia (1982)	Joint Venture : POSA 20% Australia 80% Equity : US $ 31.15 million	Mining	Coal-mimimg at MOUNT Thorley	Coking coal exported to POSCO
Pohang Steel Canada Ltd.(POSCAN) HQ.VANCOUVER, Canada (1982)	Joint Venture : POSCAN 20% Canada 80% Equity : US $ 43.2 Million	Mining	Development of Greenhills coal mine	Coking coal exported to POSCO

Sources: POSCO: Pohang and Kwanggang, May/June 1997.

Index

Abacha, Sanni, 55, 69, 113, 138, 140, 143, 162, 170, 174, 175, 177, 201, 211
Abubakar, Abdulsalami, 26, 143, 170, 173, 177, 193, 206
Africa, 10, 14, 23, 28, 30, 39, 41, 44, 47, 50-51, 69, 72, 82, 103, 129, 156, 191, 196, 199, 201, 203, 211-215, 217-219, 224
African Iron and Steel Association, 15
Agbaja iron ore deposit, 96
Ajaokuta Steel Company, 73-75, 93-94, 96-98, 101, 103, 154, 219
Akhigbe, Mike, 26
Akindele, R.A., 6
Akwa people, 107
Alternative plans of action, 207
ALCAN, 113
ALUMACO, 113
Aluminium, 111, 113-114, 143-144, 158, 218-219, 222
Amsden, Alice, 35-36, 51-52, 90, 217
Anambra Motor Manufacturing Company, 111, 114
Arap Moi, Daniel, 201
Arbed Group of Luxembourg, 178
Asia, 14, 28, 30-31, 35-36, 46, 51-52, 90, 118, 156, 173, 201, 203, 215, 217, 219-221, 224
Asian NICs, 10-11, 29-30, 35, 38-42, 44, 47, 131, 167, 196-197, 199, 201-203
Australia, 97, 102, 131, 187, 192

Austria, 70, 86, 94, 131-132, 154

Babangida, Ibrahim, 26, 136, 143, 182, 198
Baltic Engineering Group Limited, 110
Bank of Choson, 64
'Beggar thy neighbour policy', 33
Benin City, 110
Boulous Enterprises Limited, Ikeja, Lagos, 111, 171, 177, 188
Brautigam, Deborah, 30, 51, 218
Brazil, 102, 218-219
British Steel Company, 133
British trading companies, 58-59

Capacity utilisation, 125, 134, 137-139, 142-143, 145, 148, 150, 172
Capitalist accumulation, 48-49, 143, 174
Capitalist state, 22-23, 48, 174, 222, 224
Central Bank of Nigeria, 15
China, 34, 62, 66, 129-130
Chister Engineering Works Nigeria Limited, 110
Choson dynasty, 60, 62
Chung-hee, Park, 33, 80, 84, 223
Churchgate, 170
Coal, 9-11, 16, 35, 55, 71, 73, 78, 85, 89, 92-93, 97, 99-103, 122, 128, 131, 163, 186-187, 191-192, 205-206, 208, 222
Colonial administrator of Korea,

61
Colonial capitalism, 53, 67, 108
Colonial state, 17-20, 31, 56-61, 108, 116, 122, 160, 211
Colonialism, 11, 17, 29, 40, 44, 57, 61, 62, 66, 79, 115, 128, 168, 204
Committee on Steel Development, 84-85, 90, 222
'Confucian' capitalism, 197
Continental Iron and Steel Company (CISCO), 60, 69, 108, 171
COREX technology, 103
Cottage industries, 17, 53, 56, 58, 211
Crisis of intersectoral linkages, 185
Cross River Limestone Company, 96, 136

Daiichi Ginko, 64
Daewoo, 15, 117-118, 120
Delta Steel Company, 26, 76-77, 93, 224
Democracy, 84, 85, 166, 197, 202, 217, 219, 224
Deyo F., 39, 41, 52, 195, 196, 215, 218, 219
Dongguk Steel Company, 118
Donggkuk University, 6
Downstream sector of the steel industry, 60, 88, 106-109, 111, 114-117, 119, 121, 128

Early growth of steel industry, 126, 129-131, 209
Eastern Regional Government, 69, 71
ECOWAS, 213
Ekuerhare, Bright, 22, 50
Ethiopia, 23
Europe, 10, 23, 46, 54, 58, 60, 67, 69, 86, 89, 109, 113, 118, 131-132, 145, 149, 155, 173, 196, 213, 218, 222, 224
Evans, Peter, 41, 52
External environment, 210

Farming implements, 17, 58, 107, 115
Federal Office of Statistics, 157
Feldstein, M., 215
Financial problem, 175, 177-179
Finished steel products, 54, 58, 60, 68, 72, 88, 91, 93, 107-108, 110, 129, 131, 142, 148-152, 164-165, 177, 188
FIRST ALUMINIUM, 113
Ford Foundation, 6
Foreign capital, 10, 17, 20-23, 26, 41, 47-49, 53, 57, 59-60, 68, 70-71, 77, 82-83, 109, 114, 160-161, 167, 175, 198, 208, 211-212, 214
France, 12, 53, 61, 67, 70, 86, 91, 109, 126, 131, 133, 149, 160, 168, 178, 209-210

General Park, 33-34, 36-38, 48, 50, 54, 80-84, 87, 100, 116, 121, 166-167, 200
Germany, 12, 53, 58, 67, 69, 86, 91, 94, 109, 126-127, 131, 133, 144, 160, 168, 178, 190-191, 203, 209-210
Ghana, 30, 212-213
Gowon, Yakubu, 71

Hanbo steel company, 90, 104, 150, 172, 180
Hatch Associates of Canada, 25
Hatch Report of 1988, 169
Hausa/Fulani, 58
Henderson D., 196, 215
Hong Kong, 51, 69
Huan River, 29
Huntington, Samuel, 40, 220

Index

Ibeto Group of Companies, 115
Igwe, B.U.N, 24, 50, 219-220
Ihonvbere, Julius, 6
IMF, 26, 45, 49, 137, 149, 168-169, 176, 179, 184, 190, 196-198, 200, 202, 210, 212, 215, 217, 219
Import substitution strategy, 19-21, 24, 59, 67-68, 82, 107, 117, 119, 121
Inchon Iron and Steel Company, 79, 150
India, 89, 102, 112, 220
Indigenization decree, 69, 113, 114, 211
Industrial accumulation, 22-23, 48-50, 65, 163, 207, 218
Industrial minerals, 61-62, 98, 205, 213
Industrialization policy/strategy, 44-45
Integrated steel companies, 11, 18, 24, 71, 92-94, 111-112, 148, 188, 190, 207-208, 212
International Iron and Steel Institute, 126, 157
Investment code, 210-212

Kano, 20-21, 50, 223
Katsina Steel Rolling Mill, 99
Keodong-dong, Pohang, 100
Kew Metal Works, 110
KIA Motors, 15
Kiely, Ray, 51-52
Kilby, Peter, 19, 50, 89
Kobe Steel of Japan, 99, 137, 175, 214
Kolade, Bayo, 24, 50
Korea, 1, 5-6, 9-20, 28-44, 46-57, 60-67, 75, 78-92, 94, 99-106, 111, 114-122, 125-128, 130-132, 134, 145, 147-154, 156-157, 159-160, 165-168, 171-172, 174-181, 184-185, 188-193, 195-208, 211-213, 217, 220-225
Korea Development Association, 15
Korea Steel Association (KOSA), 15, 90, 104, 150, 172, 221
Korean Development Institute, 38, 90, 166, 218
Korean Development strategy, 35-36, 196, 204-205
Korean economic crisis, 195, 197, 200, 214
Korean Foundation, 6, 167
Korean war, 32, 79, 116, 205
Krugman, Paul, 196, 221
Krupp of Germany, 178
Kyongsangnam-do, 121

Late growth of steel industry, 130, 209
Latin America, 39, 41, 155, 218-220
Liberia, 213
LEE, Sing Young, 6
Lucky Goldstar, 15, 117

Machine Tools Limited, 111
Makele, Ali, 76, 89, 173
Management crisis, 159, 181-182
Manchuria, 62
Manufacturers Association of Nigeria, 15, 99, 105, 171
Mauritania, 213
Maturity stage of steel Industry, 156
Mengistu, H., 23
Metal Furniture Nigeria Limited, 109
Mini-steel rolling companies, 68
Ministry of Power and Steel, 74, 112, 122, 158, 182, 206, 219
Mistubishi, 63
Mistui, 63
Mode of surplus accumulation, 47

Modern Engineering Works Limited, 110
Monogrowth strategy, 200

National Association of Chambers of Commerce, 15
National Electric Power Authority (NEPA), 97
National Iron Ore Mining Company, 26, 73, 77, 93, 96
National Metallurgical Development Centre, 73
National Steel Council, 73-74, 187
National Steel Development Authority, 71
Natural gas, 9, 55, 92, 97, 104, 144
Niger Delta, 57-58
Niger Steel Company. Enugu, 68, 108
Nigeria, 1, 5-7, 9-32, 34, 36, 38, 40-44, 46-62, 64-94, 96-112, 114-116, 118, 120-122, 125-134, 136, 138-140, 142-152, 154, 156-196, 198-215, 217-231, 268-273
Nigeria, Private Steel Companies, 16, 24, 69, 70, 71, 77, 87-89, 95, 97, 99, 100, 104-105, 110, 135, 130, 134, 140-141, 145, 147-150, 163-164, 168, 171-172, 177, 179-180, 183-185, 187-189, 193, 208-210
Nigeria, Public Steel Companies, 16, 24-28, 56, 72-74, 76-77, 89, 95, 97-98, 100-101, 103-105, 111-112, 114, 136, 139, 147, 161-164, 169-170, 172-176, 178, 180-184, 187, 193, 206, 210
Nigerian Coal Corporation, 10, 97
Nigerian Enterprises Promotion Decree of 1972, 69
Nigerian Enterprises Promotion Decree, 69
Nigeria, Federal Office of Statistics, 157
Nigerian Indigenization Decree 1972, 69, 113, 114,
Nigerian Industrial Development Bank, 15
Nigerian Institute of International Affairs, 5, 7
Nigerian Metallurgical Society, 27, 51, 222
Nigerian Mining Corporation, 10, 15
Nigerian National Petroleum Corporation, 15
Nigerian Ports Authority, 136
Nigerian Railway Corporation, 113, 136
Nigerian Steel Policy, 209
Nippon Steel Company, 178
Nnewi, 114-115, 190
North Korea, 31, 33, 61-62, 65, 79-80

Obasanjo, Olusegun, 74, 162, 163, 170, 194
Obi/Lafia Coal deposit, 97
Obiozor, George, 6
Olukoshi, A.O., 6
Omoweh, Daniel A., 1, 7
Organisation of Economic Co-operation and Development (OECD), 45, 46, 51, 89, 167, 198, 211, 224
Osaweme, M.O., 6
Oshogbo, 24, 76-77, 94-95, 98-99, 111, 137, 139, 146, 162-163, 175, 181
Oshogbo Steel Rolling Mill, 24

Park Committee, 84-86
Park regime, 33-34, 36-38, 41, 45-46, 48, 54, 80-84, 87-88, 100, 117, 165-167
Per Capita Steel Consumption, 13, 68, 79, 108, 110, 118, 126, 140-141

Index

Peugeot Automobile Nigeria, 110, 114, 145
Pohang Steel America Corporation, 155
Pohang Steel Company (POSCO), 15, 84, 87, 100
Political economy of Korea, 18, 40, 42, 60, 80, 154, 195-196
Portuguese slave merchants, 58
POSCO Engineering and Construction Company, 153
Restructuring of Nigerian State, 208

Riva of Italy, 178
Rodelet S., 196
Russia, 62, 66, 70, 94, 170
Sammi Steel Company, 150
SCOASSEMBLY, 111, 114
Shokusan, Ginko, 64
Sino-Japanese wars, 62
Sino-Russian Communism, 31-32, 46, 78
Socialist state, 23
South Africa, 69, 103, 156
South East Asia Plan, 25, 31-32, 61, 74, 78, 82-83, 86-87, 93, 95, 101, 106, 117-119, 163, 173, 175, 205, 211, 220
South Korea. Private steel companies, 145, 177
South Korea's per capita income, 32
Spain, 61
Stainless steel, 101, 121, 148-149
State and industrialization, 11, 18, 20, 22, 36, 40, 44, 46, 67, 89, 92, 108
Steel raw materials, 5, 9-11, 14, 18, 55, 60, 68, 70, 72-73, 78, 80-81, 85-87, 91, 93, 95-98, 100, 102-103, 105, 108, 157, 163, 177, 185, 191-193, 206, 208, 213-214

Steel rolling mills, 11, 18, 24, 26-27, 59-60, 77, 87, 94-95, 99, 101, 104-105, 116, 137, 146, 150, 163-164, 170
Steel Works Limited, 110
Structural Adjustment Programme (SAP), 55, 137, 221-224

Taiwan, 30, 34, 39, 41-42, 51, 112, 114, 184, 196, 215, 217-218, 220, 222
Take-off stage, 126, 128-129, 132-133, 151, 176, 209-210
Tanzania, 23, 212, 224
Tok-Chang, Kim, 63
Total Sub-Sectoral Steel Consumption (TSSSC), 126-127
TOWER, 113
Transnational companies, 110

Union Steel Manufacturing Company, 79
Unisor-Sacilor of France, 178
United African Company (UAC), 58
United States of America, 31, 66, 78, 131-132
Universal Steel Company, 69, 108, 142, 171, 177
Upstream sector of the steel industry, 87-88, 91-93, 99-100, 102, 104-105, 109, 150, 172, 192, 208
USS-POSCO Industries, 155
USSR, 23, 34, 72, 155

Vehicle/Truck assembly plants, 114
Vietnam, 118, 131, 153, 155
Voest Alpine of Austria, 86, 94, 132, 154
Volkswagen of Nigeria Limited, 110-111, 144

VSC-POSCO Steel Corporation, 155

Wan-Son, Tae, 52
Western Region of Nigeria, 58
World Bank, 9, 12, 25-26, 29, 44, 51, 54, 71, 81, 136-137, 140, 167, 169, 196, 198, 200, 203, 210, 218, 220-221, 225
World Bank East Asian Miracle, 44, 51
World Trade Organisation (WTO), 198

Yorubas, 57
Young-Sam, Kim, 85, 198

Zimbabwe, 128, 131, 139, 212
Zimbabwe, Iron and Steel Company, 139